Time-Varying Network Optimization

TIME-VARYING NETWORK OPTIMIZATION

XIAOQIANG CAI
Department of Systems Engineering and Engineering Management
The Chinese University of Hong Kong

DAN SHA
Business School
Shanghai Institute of Foreign Trade

C. K. WONG
Department of Computer Science
The Chinese University of Hong Kong

 Springer

Xiaoqiang Cai
Dan Sha
C. K. Wong
The Chinese University of Hong Kong
Hong Kong

Library of Congress Control Number: 2007922440

ISBN-10: 0-387-71214-3 (HB)   ISBN-10: 0-387-71215-1 (e-book)
ISBN-13: 978-0-387-71214-7 (HB)   ISBN-13: 978-0-387-71215-4 (e-book)

Printed on acid-free paper.

springer.com

Contents

List of Figures

List of Tables

Preface

Network flow optimization problems may arise in a wide variety of important fields, such as transportation, telecommunication, computer networking, financial planning, logistics and supply chain management, energy systems, etc. Significant and elegant results have been achieved on the theory, algorithms, and applications, of network flow optimization in the past few decades; See, for example, the seminal books written by Ahuja, Magnanti and Orlin (1993), Bazaraa, Jarvis and Sherali (1990), Bertsekas (1998), Ford and Fulkerson (1962), Gupta (1985), Iri (1969), Jensen and Barnes (1980), Lawler (1976), and Minieka (1978).

Most network optimization problems that have been studied up to date are, however, static in nature, in the sense that it is assumed that it takes zero time to traverse any arc in a network and that all attributes of the network are constant without change at any time. Networks in the real world are, nevertheless, time-varying in essence, in which any flow must take a certain amount of time to traverse an arc and the network structure and parameters (such as arc and node capacities) may change over time. In such a problem, how to plan and control the transmission of flow becomes very important, since waiting at a node, or travelling along a particular arc with different speed, may allow one to catch the best timing along his path, and therefore achieve his overall objective, such as a minimum overall cost or a minimum travel time from the origin to the destination. There are plenty of decision making problems in practice that should be formulated as optimization models on time-varying networks. The main purpose of this monograph is to describe, within a unified framework, a series of models, propositions, and algorithms, that we have developed in this area in recent years. Additional references and discussions on relevant problems and studies that have appeared in the literature will also be provided.

This monograph consists of eight chapters, in which we formulate and study respectively the shortest path problem, minimum-spanning tree problem, maximum flow problem, minimum cost flow problem, maximum capacity path problem, quickest path problem, multi-criteria problem, and generalized flow problem (the time-varying travelling salesman problem and the Chinese postman problem will also be considered in a chapter together with the time-varying generalized problem). While these topics will be described all within the framework of time-varying networks, our plan is to make each chapter relatively self-contained so that they can be read separately. It is hoped that this book is useful for researchers, practitioners, and graduate students and senior undergraduates, as a unified reference and textbook on time-varying network optimization. While we describe in this book only the structure of the algorithms, we have developed the software that implements the algorithms, which is available for academic study purpose upon request.

The publication of this book would not be possible without the help and generous support of many people and organizations. First, we would like to express our sincere gratitude to Professor Fred Hillier, the Editor of the book series, for his encouragement and guidance. We are very indebted to Gary Folven, Publisher of Springer, for his kind invitation, continued support, and patience, to allow us to complete this book project. We also thank Carolyn Ford of Springer, for her careful reading and processing of our manuscript. Many colleagues and students have kindly provided us with invaluable comments and suggestions in various occasions such as research seminars and conferences. A number of our students have contributed to the literature survey and software development of the algorithms presented in this book, including Chyrel Teo, Wenting Hou, just to name a few. Part of our research effort leading to the results included in this book was financially supported by the Research Grants Council of Hong Kong under Project Nos CUHK4135/97E, CUHK 4170/03E, and N_CUHK442/05, and the National Science Foundation of China (NSFC) Research Fund Nos. 70329001 and 70518002.

Last but not the least, the greatest gratitude must go to our families, for their support and patience over the many days and nights during the writing of this book.

XIAOQIANG CAI, DAN SHA, AND C.K. WONG

Chapter 1

TIME-VARYING SHORTEST PATH PROBLEMS

1.   Introduction

The static version of the shortest path problem assumes that zero time is required to traverse any path in the network, where all problem parameters are not changed at all over time. This problem is the most fundamental one in network optimization, which has been widely studied in the literature. An extension of the static model is the problem of finding a shortest path subject to some constraints involving transit times. An example is the problem where a transit time is required to traverse an arc, subject to the constraint that the total time to traverse a path cannot be greater than any given amount T. This class of problems has been studied by Aneja (1978); Hassan (1992); Joksch (1966); Skiscim et al (1989); Witzgall et al (1965), and is known to be NP-complete; see Ahuja et al (1993); Handler et al (1980).

We will consider, in this chapter, the time-varying version of the problem. A general model will be addressed, in which the transit time and the cost to traverse an arc are varying over time, which depend upon the departure time at the beginning vertex of the arc. Moreover, waiting at a vertex is allowed, at a waiting cost; and speedup on an arc is also possible, at a speedup cost. The problem is to find an optimal path to travel from a source vertex s to another vertex x, so that the total cost of the path is minimized subject to the constraint that the total travel time of the path is at most T, where T is a given integer. In addition to the determination of a path to connect the two vertices s and x, the waiting times at all vertices and the speedups on all arcs along the path should also be decision variables to be determined. We call this the *"time-varying shortest path"* (TVSP) problem.

Desrosiers et al (1986) have considered a *shortest path problem with time windows*. In their model, it is assumed that each arc has a constant transit time and a constant transit cost, and each vertex x_i can only be visited during a time period $[a_i, b_i]$, which is called the *time window*. Given a time limit T, they want to find the shortest path from the source vertex to another vertex under the time window constraint. Clearly, by setting the waiting cost at x_i at any time $t < a_i$ to be infinite, and the transit cost to depart from x_i at any time $t < a_i$ or $t > b_i$ to be infinite, one can see that this problem becomes a special case of the TVSP problem described above.

The TVSP model has many applications. One example is the data transmission problem. Suppose that a data packet has to be sent between two specified nodes in a network as cheaply as possible within a time limit T. As the transit time and the cost needed to send the packet on an arc vary over different periods, there exists an optimal departure time to traverse an arc. Thus, an optimal solution to the problem should not only provide the best path connecting the two nodes, but also specify the optimal duration for the data packet to stay at each node to wait for the best departure time. Another example is the freight transport problem. Suppose that some freight is to be sent from a source to a sink in a network before a deadline T. Between two neighboring cities, several types of freight services are available, which, however, have different costs and transport times, depending upon the seasons. An optimal solution should thus specify the route as well as the waiting times of the freight at each city so that the overall cost is minimum while the freight arrives at the destination no later than the deadline T.

This chapter will be organized as follows: In Section 2, we introduce the necessary concepts and the basic model. The optimality properties and the NP-completeness of the problem will be discussed in Section 3. Then, Section 4 is devoted to problems with arbitrary waiting times, zero waiting times, and bounded waiting times, respectively, all under the assumption of strictly positive times. The results are generalized to the case with nonnegative transit times in Section 5. The model with speedups will be considered in Section 6. Finally, some additional references and comments are given in Section 7.

2.   Concepts and problem formulation

Let $N(V, A, b, c)$ be a network without parallel arcs and self-loops, where V is the set of vertices, A is the set of arcs, $b(x, y, t)$ is the *transit time* needed to traverse an arc (x, y). This transit time is dependent on the discrete values of the time $t = 0, 1, ..., T$, where $T \geq 0$ is the maximum allowable time to travel from the origin vertex s to the target vertex x. Let $c(x, y, t)$ be the *transit cost* of an arc $(x, y) \in A$. This transit cost is also dependent on the discrete value of the time $t = 0, 1, ..., T$. Both the transit time $b(x, y, t)$ and the transit cost $c(x, y, t)$ are functions of the departure time t at the beginning vertex x of the arc (x, y). Moreover, let $c(x, t)$ be the *waiting cost* if waiting takes place at vertex x during the time period from t to $t + 1$, which is a function of time t. We assume that both the transit cost and the waiting cost are arbitrary integers, while the transit time b is a nonnegative integer. Moreover, we let $n = |V|$ and $m = |A|$.

We also assume that there is a unique source vertex in N, denoted by s. If the network has more than one source vertex s_i, $i = 1, 2, \cdots, I_s$, one can introduce a dummy source vertex s^* and I_s arcs (s^*, s_i) to connect it with each source vertices s_i, so that the network contains only one source vertex.

If the network contains parallel arcs or self-loops, we can convert it into one with no parallel arcs and self-loops, by inserting an artificial vertex into those arcs or self-loops. The following example illustrates such a conversion.

Example 1.1

Figure 1.1. Convert a network into one without parallel arcs and loops

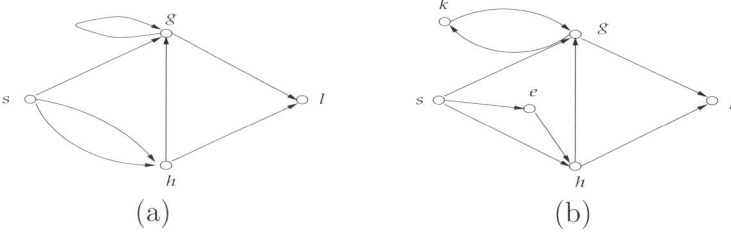

(a)   (b)

Given a network N as shown in Figure 1.1(a), there are two parallel arcs (s, h) and a loop (g, g) in N. We insert an artificial vertex e into one of the parallel arcs (s, h), and another vertex k into the loop (g, g) (see Figure 1.1(b)). Furthermore, for any time t, let

$$b(s, e, t) = 0, b(e, h, t) = b(s, h, t), b(g, k, t) = 0, b(k, g, t) = b(g, g, t),$$

$$c(s, e, t) = 0, c(e, h, t) = c(s, h, t), c(g, k, t) = 0, c(k, g, t) = c(g, g, t),$$

$$c(e, t) = c(k, t) = 0,$$

and all other numbers maintain unchanged. Clearly, the new network has no parallel arcs and self-loops, and finding the shortest path in the new network is equivalent to finding the shortest path in the original one.

We need the following concepts and notation.

Definition 1.1 *A waiting time $w(x)$ at vertex x is a nonnegative integer, and u_x is its upper bound.*

Definition 1.2 *Let $P(x_1, ..., x_r)$ be a path from x_1 to x_r. The arrival time of a vertex x_i on P is defined as $\alpha(x_i)$ such that $\alpha(x_1) = t_0 \geq 0$ (for the source vertex s, we let $\alpha(s) = 0$), and*

$$\alpha(x_i) = \alpha(x_{i-1}) + w(x_{i-1}) + b(x_{i-1}, x_i, \tau(x_{i-1})), \qquad for\ i = 2, ..., r,$$

where $\tau(x_i)$, the departure time of a vertex x_i on P, is defined as

$$\tau(x_i) = \alpha(x_i) + w(x_i), \qquad for\ i = 1, ..., r - 1.$$

Clearly, when we consider a path in a time-varying network, we have to consider not only the graph structure of the path, but also the relevant arrival times, departure times, and waiting times, at all its vertices. The changes of the transit times by speedups on all its arcs should also be considered. To simplify our discussion, we will, nevertheless, leave the model with speedups to be introduced in Section 6. In what follows, we introduce the concept of a dynamic path, with $\alpha(x_i)$, $w(x_i)$, and $\tau(x_i)$ specified, under the assumption that the transit times $b(x, y, t)$ cannot be shortened.

Definition 1.3 *$P = (x_1, ..., x_r)$ is said to be a dynamic path from x_1 to x_r, if all the $\alpha(x_i)$, $w(x_i)$, and $\tau(x_i)$ on the path are specified. Furthermore, the time of P is defined as $\alpha(x_r) + w(x_r) - \alpha(x_1)$. A path is said to have time at most t, if its time is less than or equal to t. Specifically, a path is said to have time exactly t, if its time equals t.*

For any dynamic path $P = (s, ..., x)$, the time of P is $\alpha(x) + w(x)$, since we can assume, without loss of generality, that the arrival time at s, $\alpha(s)$, is 0.

Definition 1.4 *Let $P = (x_1, ..., x_r)$ be a dynamic path from x_1 to x_r. Let $\zeta(x_1) = \sum_{t'=0}^{w(x_1)-1} c(x_1, t' + \alpha(x_1))$ and define recursively*

$$\zeta(x_i) = \zeta(x_{i-1}) + c(x_{i-1}, x_i, \tau(x_{i-1})) + \sum_{t'=0}^{w(x_i)-1} c(x_i, t' + \alpha(x_i))$$

for $i = 2, ..., r$. The cost (or length) of P, $\zeta(P)$, is defined as $\zeta(x_r)$.

For completeness, we adopt the following convention:

Definition 1.5 *For vertices $x, y \in V$ and a given number $t \leq T$, the cost of a shortest path from x to y within time t is said to be ∞ if*
(i) there does not exist any path from x to y, or
(ii) all paths from x to y have times greater than t.

3.   Properties and NP-completeness

The time-varying structure of the network induces a number of interesting phenomena. For example, some important properties that exist for the static shortest path problem have, however, no longer held in the time-varying model.

It is an important property for the static shortest path problem that, if $P(x_1, ..., x_r)$ is a shortest path from x_1 to x_r, then for every vertex x_i ($i = 2, ..., r - 1$), the subpath $P'(x_1, ..., x_i)$ is a shortest path from x_1 to x_i. This property seems to be straightforward for a shortest path problem. Nevertheless, it is no longer valid for the time-varying model. Let us examine the example as shown in Figure 1.2, where $T = 8$, and

$$b(s, f, 0) = 2, b(f, g, 2) = 1, b(s, g, 0) = 5, b(g, h, 5) = 3,$$

$$c(s, f, 0) = 1, c(f, g, 2) = 1, c(s, g, 0) = 3, \text{ and } c(g, h, t) = 2 \text{ for } t \geq 4.$$

All other b and c are equal to ∞. Moreover, $u_g = 0$, i.e., waiting at the vertex g is now allowed.

Figure 1.2.   A subpath of the shortest path is not a shortest path

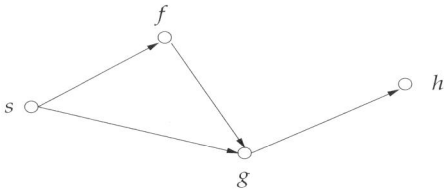

Clearly, the shortest (cheapest) path from s to h of time at most 8 is $P(s, g, h)$, with $\tau(s) = 0$, $\alpha(g) = \tau(g) = 5$, and $\alpha(h) = 8$. The cost of the path $\zeta(P(s, h))$ is 5. Its subpath $P'(s, g)$ has a cost 3, which is, however, not the shortest path from s to g. The shortest path from s to g should be $P(s, f, g)$, with $\tau(s) = 0$, $\alpha(f) = \tau(f) = 2$, $\alpha(g) = 3$ and $\zeta(P(s, f, g)) = 2$.

The subpath $P'(s, g)$ is not the shortest one, but it allows one to depart from g at time 5 and therefore catch the feasible time to traverse the arc (g, h).

The example above indicates that the property for the static problem cannot be simply extended to the time-varying model. Under some additional restrictions, the property may continue to hold; see below.

Property 1.1 *Suppose that the path $P(x_1, ..., x_r)$ is a shortest dynamic path from x_1 to x_r of time exactly t, then for each vertex x_i ($i = 1, 2, ..., r$), the subpath $P'(x_1, ..., x_i)$ is a shortest path from x_1 to x_i of time exactly t', under the restriction that $\alpha(x_i) \leq t' \leq \tau(x_i)$.*

Another important property for the static shortest path problem indicates that, if all arc costs are positive numbers, then any shortest path must be a simple path (A simple path is defined as one in which each vertex appears only once). Again, this property has no longer held in the time-varying model. Let us look at the following example, where the network is given in Figure 1.3. Moreover,

$$b(s, g, 0) = 1, b(g, q, 1) = 1, b(q, h, 2) = 1,$$

$$b(h, g, 3) = 1, b(g, q, 4) = 1, b(q, i, 5) = 1,$$

while all other b's are ∞. The transit costs $c(x, y, t) = 1$, for the arc $(x, y) \in A$ and $t \geq 0$, and the waiting costs $c(x, t) = \infty$ for the vertex $x \in V$ and $t \geq 0$. The deadline $T = 6$.

Figure 1.3. A shortest path is not a simple path

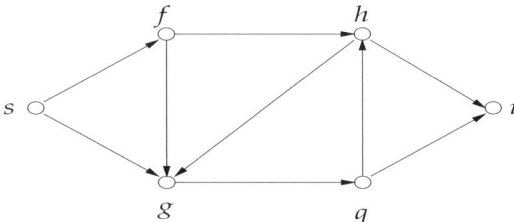

It is easy to verify that there is only one shortest dynamic path from s to i within time $T = 6$, which is $P = (s, g, q, h, g, q, i)$, with the total cost being 6. This is the optimal path, in which the vertices g and q appear, however, twice, and therefore it is not a simple path.

Again, we need some additional restrictions to make the property valid in a time-varying network. For instance, if $c(x, y, t) > 0$, $c(x, t) = 0$, and $u_x = \infty$, for all x, y and t, then the property will remain true. This result can be established by the following observation: If there is a shortest dynamic path $P(s, ..., x', ..., x', ..., x)$, where x' is visited twice in path P at time t_1 and t_2 ($t_1 < t_2$), then we can wait at x' from time

t_1 to t_2 and skip those vertices between two x's. Let P' denote the new path after deleting the section $(x', ..., x')$ of P. Then, it is clear that the cost of P' is not greater than that of P. This result is given below.

Property 1.2 *There exists a simple shortest path in a time-varying network N, if waiting at any vertex of N is allowed subject to no constraint and causing no cost, and all transit costs are positive.*

There are other scenarios where the time-varying model creates new issues, which we leave to the readers who are interested to analyze and investigate. In what follows we examine the complexity of the time-varying model, TVSP.

It has been well-known that the static shortest path problem is polynomially solvable. The time-varying problem, TVSP, is however NP-complete, as we will show below. First, let us introduce the decision version of TVSP, which can be described as follows.

TVSP Given a time-varying network $N(V, A, b, c)$ and an integer K, does there exist a dynamic path P from s to ρ within the time T, such that $\zeta(P) \leq K$?

We will show that the Knapsack problem is reducible to TVSP. Knapsack is a well-known NP-complete problem (Garey and Johnson 1979), which can be stated as:

Knapsack Given a finite set Q of elements, each having a value v_i and a size w_i, and two integers v^* and w^*, does there exist a subset $S \subseteq Q$ such that $\sum_{i \in S} v_i \geq v^*$ and $\sum_{i \in S} w_i \leq w^*$?

The following theorem establishes the NP-completeness of the TVSP problem.

Theorem 1.1 *The TVSP problem is NP-complete in the ordinary sense.*

Proof. For any given instance of Knapsack, we can construct a network N as shown in Figure 1.4. Let $B = \sum v_i$, $K = nB - v^*$, $T = n + w^* + 1$, and

$$b(x_{i-1}, x'_i, t) = 1, b(x'_i, x_i, t) = w_i, 1 \leq i \leq n, 0 \leq t \leq T$$

$$c(x_{i-1}, x'_i, t) = 0, c(x'_i, x_i, t) = B - v_i, 1 \leq i \leq n, 0 \leq t \leq T$$

$$b(x_{i-1}, x_i, t) = 1, c(x_{i-1}, x_i, t) = B, 1 \leq i \leq n, 0 \leq t \leq T$$

$$b(x_n, x_{n+1}, t) = 1, 0 \leq t \leq T$$

$$c(x_n, x_{n+1}, n + w^*) = 0, c(x_n, x_{n+1}, t) = \infty, 0 \leq t \leq T, t \neq n + w^*$$

Figure 1.4.   A time-varying network constructed from Knapsack

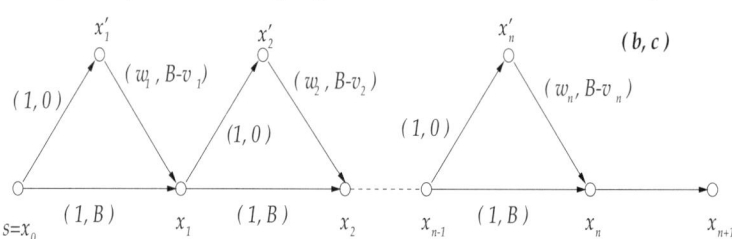

Figure 1.4.   A time-varying network constructed from Knapsack

We now prove that a "yes" answer to Knapsack is equivalent to a "yes" to the decision version of TVSP.

If Knapsack has a set $S \subseteq Q$ such that $\sum_{i \in S} v_i \geq v^*$ and $\sum_{i \in S} w_i \leq w^*$, then we can obtain a path $P(x_0, x_{n+1})$ in the following way: starting from x_0 at time zero, for each i, if $i \in S$, then traverse arcs (x_{i-1}, x'_i) and (x'_i, x_i); if $i \notin S$, then traverse arc (x_{i-1}, x_i). At last, traverse arc (x_n, x_{n+1}). Let $w(x_i) = 0$ $(1 \leq i \leq n-1)$ and $w(x_n) = n + w^* - \alpha(x_n)$. Notice that, since $\sum_{i \in S} w_i \leq w^*$, we have $\alpha(x_n) = \sum_{i \in S}(w_i + 1) + \sum_{i \notin S} 1 = n + \sum_{i \in S} w_i \leq n + w^*$, and $\alpha(x_{n+1}) \leq n + w^* + 1 = T$, where $\alpha(x_n)$ and $\alpha(x_{n+1})$ are the arrival times of P at vertices x_n and x_{n+1}, respectively. Clearly, P is a dynamic path from s to ρ within time duration T. Moreover, since $\sum_{i \in S} v_i \geq v^*$, i.e., $-\sum_{i \in S} v_i \leq -v^*$, we have $\zeta(P) = \sum_{i \notin S} B + \sum_{i \in S}(B - v_i) = nB - \sum_{i \in S} v_i \leq nB - v^* = K$.

Now, suppose that TVSP has a dynamic path P of time at most T with $\zeta(P) \leq K$. Let $S = \{i | (x'_i, x_i) \in A(P), 1 \leq i \leq n\}$, where $A(P)$ is the arc set of P. We have $\zeta(P) = \sum_{i \in S} c(x'_i, x_i, \tau(x'_i)) + \sum_{i \notin S} B = \sum_{i \in S}(B - v_i) + \sum_{i \notin S} B = nB - \sum_{i \in S} v_i \leq nB - v^* = K$, which implies $\sum_{i \in S} v_i \geq v^*$. On the other hand, since P is a dynamic path within time T, by the construction of the network N, we must have $\alpha(x_n) = \sum_{i \in S}(w_i + 1) + \sum_{i \notin S} 1 = n + \sum_{i \in S} w_i \leq n + w^*$, where $\alpha(x_n)$ is the arrival time of P at x_n. Therefore, $\sum_{i \in S} w_i \leq w^*$.

The analysis above show that TVSP can be reduced from Knapsack and thus it is NP-complete. On the other hand, the optimal solution of TVSP can be found by an algorithm in pseudopolynomial time (see Section 4 below) and hence it is NP-complete in the ordinary sense (cf. Garey and Johnson 1979). This completes the proof.   □

4.   Algorithms

As we have mentioned before, the waiting time at a vertex is a decision variable that should be considered in a time-varying network model. Because of the time-varying nature of the network, departing from a vertex too early may not be desirable. In other words, waiting at a vertex to postpone the departure from the vertex may be a better decision.

Waiting at a vertex is often subject to constraints. In order to highlight the basic ideas of the algorithms to be presented below, we will consider, in this section, only the constraint that the waiting time at a vertex x, $w(x)$, is subject to an upper bound u_x.

Corresponding to the situations with $w(x) = \infty$, $w(x) = 0$, and $w(x) = u_x$ with u_x being a finite positive number, we have the TVSP problem with arbitrary waiting times (TVSP-AWT), the TVSP problem with zero waiting times (TVSP-ZWT), and the TVSP problem with bounded waiting times (TVSP-BWT), respectively. In this section, we will introduce an algorithm for each of these problems, under the assumption that all $b(x, y, t)$ are positive integers. The case with some $b(x, y, t)$ equal to zero will be discussed in Section 5.

As we have indicated, the transit time plays the most important role in searching for the solutions in a time-varying network. Because of the positive time requirement to reach a vertex, the search for an optimum can be limited to a local region (in other words, vertices that would be impossible to reach at a time t could be ignored before the time t). This makes it possible for us to develop efficient dynamic programming algorithms to construct a solution with a forward pass with respect to the time t. This will be elaborated below. The algorithms described in this section mainly come from Cai, Kloks, and Wong (1997).

4.1   Waiting at any vertex is arbitrarily allowed

We now consider the problem TVSP-AWT. First, let us introduce the notation $d_a(y, t)$.

Definition 1.6 *Define $d_a(y, t)$ as the cost of a shortest path from s to y of time exactly t, where waiting at any vertex is not restricted.*

The following lemma gives us a recursive relation to compute $d_a(x, t)$. Note that the optimal waiting times can be obtained implicitly by the recursive computations. This will be further elaborated in Remark 1.1.

Lemma 1.1 $d_a(s, 0) = 0$, and $d_a(y, 0) = \infty$ for all $y \neq s$. For $t > 0$, we have

$$d_a(y, t) = \min \Big\{ d_a(y, t - 1) + c(y, t - 1),$$

$$\min_{(x,y) \in A} \min_{\{u \mid u + b(x,y,u) = t\}} \{ d_a(x, u) + c(x, y, u) \} \Big\}.$$

Proof. It is easy to see that $d_a(s, 0) = 0$, and $d_a(y, 0) = \infty$ for all $y \neq s$, since all transit times are positive.

Now prove the formula by induction. Consider $t = 1$. The only vertices for which there can exist a path of time exactly one are s and neighbors of s. For $y = s$, the formula clearly holds. Assume that $y \neq s$. Consider first the case where $(s, y) \subset A$ and $b(s, y, 0) = 1$. In this case, the formula holds with $d_a(y, t) = d_a(x, u) + c(x, y, u)$, where $u = 0$ and $x = s$. In any other cases, the formula holds with $d_a(y, t) = d_a(y, t - 1) + c(y, t - 1) = \infty$ as there is no feasible solution for a path from s to y of time $t = 1$.

Assume that the formula is correct for all $t' < t$. Consider a vertex y. First, we prove the claim that there exists a path of time exactly t and cost $d_a(y, t)$.

If $d_a(y, t) = \infty$, there is nothing to prove. So, assume that $d_a(y, t)$ is finite. If $d_a(y, t) = d_a(y, t - 1) + c(y, t - 1)$, then, by induction, there is a path from s to y of cost $d_a(y, t - 1)$. The time of the path is $t - 1$. By waiting at y one unit of time more, the path has time exactly t.

Assume that $d_a(y, t) = d_a(x, u) + c(x, y, u)$ for some x such that $(x, y) \in A$ and some u such that $u + b(x, y, u) = t$. Since $b(x, y, u) > 0$, we have $u < t$ and, therefore, by induction, we know there must exist a path $P'(s = x_1, ..., x_r = x)$ from s to x of time exactly u and cost $d_a(x, u)$. Hence, there are waiting times $w(x_i)$ at x_i such that the time of the path P' with these waiting times is u. we extend the path to vertex y, obtaining a path P with the given waiting times and with waiting time zero at y. The time of P is exactly t, since $u + b(x, y, u) = t$. The cost of P with these waiting times is $d_a(x, u) + c(x, y, u) = d_a(y, t)$. This proves the claim.

We now prove that $d_a(y, t)$ is the cost of a shortest path from s to y with time exactly t. Let $P(s = x_1, ..., x_r = y)$ be a shortest path from s to y of time exactly t, and let $w(x_i)$ be the waiting time at x_i $(i = 1, ..., r)$. If $w(x_r) > 0$, then let P' be a path same as P but waiting at x_r for $w(x_r) - 1$. Clearly, P' is a shortest path from s to x_r of time exactly $t - 1$. By induction, $d_a(y, t - 1) \leq \zeta(P')$. Therefore, we have $\zeta(P) = \zeta(P') + c(x_r, t - 1) \geq d_a(x_r, t - 1) + c(x_r, t - 1) = d_a(x_r, t)$. On the other hand, we have $\zeta(P) \leq d_a(x_r, t)$ since P is a shortest path from s to x_r of time exactly t. Thus, $\zeta(P) = d_a(x_r, t)$. Now, we consider the case $w(x_r) = 0$. Let x be the predecessor of y on this path. Let u be the time of the subpath P' (with the waiting times) from s to x, and let $\zeta(P')$ be the cost of P'. By definition, $u + b(x, y, u) + w(x_r) = u + b(x, y, u) = t$, implying that $u < t$ since $b(x, y, u) > 0$. Thus, by induction, $\zeta(P') \geq d_a(x, u)$. By the definition, the cost of P is $\zeta(P) = \zeta(P') + c(x, y, u)) \geq d_a(x, u) + c(x, y, u)$. Again, according to the formula, we have $d_a(y, t) = d_a(x, u) + c(x, y, u) \leq \zeta(P)$. In both cases, we must have $\zeta(P) = d_a(y, t)$, since P is a shortest path and since there exists a path achieving $d_a(y, t)$, as we showed above. This completes the proof. □

Definition 1.7 *Define $d_a^*(x)$ as the cost of a shortest dynamic path from s to x of time at most T, where waiting at vertex is arbitrary allowed.*

Lemma 1.2
$$d_a^*(x) = \min_{0 \leq t \leq T} d_a(x, t)$$

Proof. Consider a shortest feasible path P of time at most T. Let $t \leq T$ be the time of P, with waiting time zero at x. Then, $d_a^*(x) = d_a(x, t)$. □

Definition 1.8 *For every arc $(x, y) \in A$ and for $t = 0, ..., T$, let*

$$R_a(x, y, t) = \min_{\{u | u + b(x,y,u) = t\}} \{d_a(x, u) + c(x, y, u)\}$$

We adopt the convention that $R_a(x, y, t) = \infty$ whenever $\{u | u + b(x, y, u) = t\} = \emptyset$.

The result below follows directly from Lemma 1.1:

Corollary 1.1
$$d_a(y, t) = \min\{d_a(y, t - 1) + c(y, t - 1), \min_{\{x | (x,y) \in A\}} R_a(x, y, t)\}.$$

Corollary 1.1 indicates that when $d_a(y, t)$ is to be updated we have to know $R_a(x, y, t)$ for all $(x, y) \in A$. Given t and (x, y), $R_a(x, y, t)$ could be evaluated by a naive approach of enumerating $0 \leq u \leq t$ to find those satisfying $u + b(x, y, u) = t$, according to Definition 1.8. This would, however, require a worse-case running time of $O(T)$ for every t. Clearly, we need some mechanism to make the evaluation of $R_a(x, y, t)$ efficiently. Our idea in the algorithm below is to first sort the values of $u + b(x, y, u)$ for all $u = 1, 2, ..., T$ and all arcs $(x, y) \in A$, before the recursive relation as given in Lemma 1.1 is applied to compute $d_a(y, t)$ for all $y \in V$ and $t = 1, 2, ..., T$.

We describe the algorithm as below:

> **Algorithm TVSP-AWT**
> **Begin**
> Initialize $d_a(s, 0) = 0$ and $\forall_{x \neq s} d_a(x, 0) = \infty$;
> Sort all values $u + b(x, y, u)$ for $u = 0, 1, ..., T$ and for all arcs $(x, y) \in A$;
>   **For** $t = 1, ..., T$ **do**
>   **For** every arc $(x, y) \in A$ **do** $R_a(x, y, t) := \infty$;

> **For** all arcs $(x, y) \in A$ and all u such that $u + b(x, y, u) = t$ **do**
> $R_a(x, y, t) := \min\{R_a(x, y, t), d_a(x, u) + c(x, y, u)\};$
> **For** every y **do**
> $d_a(y, t) := \min\{d_a(y, t \quad 1) \mid c(y, t \quad 1), \min_{(x,y) \in A} R_a(x, y, t)\};$
> **For** every y **do** $d_a^*(y) := \min_{0 \leq t \leq T} d_a(y, t);$
> **End**.

We now consider a simple example.

Example 1.2

Assume that there is a network as shown in Figure 1.5, where the two elements in the square brackets along each arc (x, y) represent the transit time $b(x, y, u)$ and the cost $c(x, y, u)$ of the arc, respectively. The problem is to find a shortest path connecting the source node s and the sink node i such that the time of the path is at most $T = 12$, where $c(x, t) = 0$ for all x and all time t.

Figure 1.5.   Example 1.2

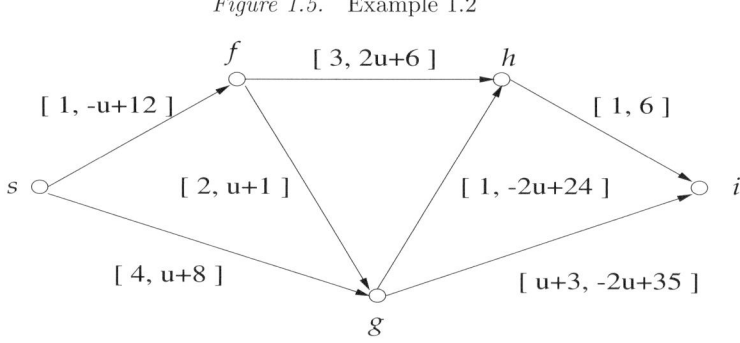

Applying Algorithm TVSP-AWT, one may obtain the results in Table 1.1. Thus, when $T = 12$, the cost of the shortest path connecting s and i is $d_a(i, 12) = 18$. By a backtracking procedure, it is easy to find that the shortest path is $P^*(s, g, h, i)$, where the departure times at the vertices s, g, and h are, respectively, 0, 10, and 11, while the arrival times at the vertices g, h, and i are, respectively, 4, 11, and 12. There is a waiting time 6 at the vertex g to achieve the minimal cost 18.

Remark 1.1 In general, the algorithm TVSP-AWT computes the cost $d_a(x, t)$ of the shortest path from the source s to a vertex x with time at most T. Let the shortest path be $P^*(s = x_1^*, ..., x_i^*, x_j^*, ..., x_r^* = x)$. Note that this path and the optimal departure time at each vertex x_i^* on the path can be identified by a standard backtracking procedure of

Table 1.1. Calculation of shortest path

t	$d_a(s,t)$	$d_a(f,t)$	$d_a(g,t)$	$d_a(h,t)$	$d_a(i,t)$
0	0	∞	∞	∞	∞
1	0	12	∞	∞	∞
2	0	11	∞	∞	∞
3	0	10	14	∞	∞
4	0	9	8	20	∞
5	0	8	8	20	26
6	0	7	8	20	26
7	0	6	8	20	26
8	0	5	8	18	26
9	0	4	8	16	24
10	0	3	8	14	22
11	0	2	8	12	20
12	0	1	8	10	18

dynamic programming. Then, the waiting times at the vertices on P^* can be obtained using the departure times. For example, if $\tau(x_i^*)$ and $\tau(x_j^*)$ are the optimal departure times at two vertices x_i^* and x_j^* of an arc (x_i^*, x_j^*) on P^*, then the optimal waiting time at the vertex x^* is $w(x_j^*) = \tau(x_j^*) - \tau(x_i^*) - b(x_i^*, x_j^*, \tau(x_i^*))$.

Lemma 1.3 *The algorithm TVSP-AWT can be implemented such that it runs in $O(T(m+n))$ time.*

Proof. It is easy to check that the initialization can be done in $O(Tn)$ time.

For the sorting in step 2, we can use bucketsort, with T buckets. Since there are Tm values to be sorted, this step can then be performed in $O(Tm)$ time.

Since the values $u + b(x, y, u)$ are now sorted, the overall time needed to update the values $R_a(x, y, t)$ can be done in $O(Tm)$ time.

Finally, the overall time to update the values $d_a(y, t)$ in the last line is proportional to T times the number of arcs, i.e., $O(Tm)$.

It follows that the running time of the algorithm is bounded by $O(T(m+n))$. $\qquad\square$

From Lemma 1.1 and Corollary 1.1, one can easily see that, after the termination of the algorithm TVSP-AWT, each computed value $d_a(x, t)$ is the cost of a shortest path from s to x of time at most t. This together with Lemma 1.3, gives us

Theorem 1.2 *The TVSP-AWT problem with positive transit times can be optimally solved in $O(T(m+n))$ time.*

4.2   Waiting at any vertex is prohibited

We now consider the problem TVSP-ZWT, in which no waiting times are allowed at any vertices. Recall Definitions 1.2, 1.3, and 1.4 on departure times at vertices, time of path, and cost of path. Because the waiting times in these definitions should be set to zero for the TVSP-ZWT problem, we need not consider the waiting cost.

Definition 1.9 *Define $d_z(y,t)$ as the cost of a shortest path from s to y of time exactly t. If such a path does not exist, then $d_z(y,t) = \infty$.*

Following similar arguments in the proof for Lemma 1.1, one can show

Lemma 1.4 $d_z(s,0) = 0$ *and* $d_z(y,0) = \infty$ *for all* $y \neq s$. *For* $t > 0$, *we have*

$$d_z(y,t) = \min_{\{x|(x,y)\in A\}} \min_{\{u|u+b(x,y,u)=t\}} \{d_z(x,u) + c(x,y,u)\}.$$

Let us now further introduce the following definition:

Definition 1.10 *For each arc $(x,y) \in A$ and each $1 \leq t \leq T$, define*

$$R_z(x,y,t) = \min_{\{u|u+b(x,y,u)=t\}} \{d_z(x,u) + c(x,y,u)\}$$

and adopt the convention that $R_z(x,y,t) = \infty$ whenever the set $\{u|u + b(x,y,u) = t\}$ is empty.

The result below follows directly from Lemma 1.4:

Corollary 1.2 *For $1 \leq t \leq T$, and for each vertex y,*

$$d_z(y,t) = \min_{\{x|(x,y)\in A\}} R_z(x,y,t)$$

Let $d_z^*(x)$ be the cost of a shortest path from s to x of time at most T. Clearly, we have $d_z^*(x) = \min_{0 \leq t \leq T} d_z(x,t)$. Now we describe the algorithm for solving the TVSP-ZWT problem as below:

> **Algorithm TVSP-ZWT**
> **Begin**
> Initialize $d_z(s,0) = 0$ and $\forall_{x \neq s} d_z(x,0) = \infty$;
> Sort all values $u + b(x,y,u)$ for $u = 0, 1, ..., T$ and for all arcs $(x,y) \in A$;

For $t = 1, ..., T$ **do**
  For every arc $(x, y) \in A$ **do** $R_z(x, y, t) := \infty$;
  For all arcs $(x, y) \in A$ and all u such that $u + b(x, y, u) = t$
do
$$R_z(x, y, t) := \min\{R_z(x, y, t), d_z(x, u) + c(x, y, u)\};$$
  For every y **do** $d_z(y, t) := \min_{(x,y) \in A} R_z(x, y, t)$;
  For every y **do** $d_z^*(y) := \min_{0 \le t \le T} d_z(y, t)$;
  End.

The following lemma gives the worst-case running time of the algorithm TVSP-ZWT. The proof of the lemma is similar to that for Lemma 1.3 and is omitted here.

Lemma 1.5 *The algorithm TVSP-ZWT can be implemented such that it runs in $O(T(m + n))$ time.*

Theorem 1.3 *The TVSP-ZWT problem can be optimally solved in $O(T(m + n))$ time.*

4.3 Waiting time is subject to an upper bound

We now consider the problem TVSP-BWT, where waiting at a vertex is allowed, but is constrained by an upper-bound u_x. Clearly, we may assume that $u_x \le T$ for all $x \in V$.

Definition 1.11 $d_b(x, t)$ *is the cost of a shortest feasible path from s to y of time exactly t and with waiting time zero at x, subject to the constraint that the waiting time at any other vertex y on the path is not greater than u_y. If such a feasible path does not exist, then $d_b(x, t) = \infty$.*

Lemma 1.6 $d_b(s, 0) = 0$ *and* $d_b(x, 0) = \infty$ *for all $x \neq s$. For $t > 0$, we have*

$$d_b(y, t) =$$

$$\min_{\{x \mid (x,y) \in A\}} \min_{(u_A, u_D) \in \mathcal{F}(x,y,t)} \left\{ d_b(x, u_A) + c(x, y, u_D) + \sum_{t'=0}^{u_D - u_A - 1} c(x, t' + u_A) \right\}$$

where $\mathcal{F}(x, y, t) = \{(u_A, u_D) \mid u_D + b(x, y, u_D) = t \wedge 0 \le u_D - u_A \le u_x\}$, *and u_A and u_D are the arrival time and the departure time at x respectively.*

Proof. It is easy to see that $d_b(s, 0) = 0$ and $d_b(y, 0) = \infty$ for all $y \neq s$, since all transit times are positive. Thus, in the following, we need only examine $t > 0$.

We prove the formula by induction. Consider $t = 1$. The only vertices for which there exists a feasible path of time one are neighbors of s. Assume that y is a neighbor of s. We must have $b(s, y, 0) = 1$ and all waiting times must be zero. In that case, the formula holds with $u_A = u_D = 0$ and $x = s$.

Assume that the formula is correct for all $t' < t$. Consider a vertex y. First, let us prove the claim that there exists a feasible path of time t and cost $d_b(y, t)$, with waiting time zero at vertex y.

If $d_b(y, t) = \infty$, there is nothing to prove. So assume that $d_b(y, t)$ is finite. Assume that $d_b(y, t) = d_b(x, u_A) + c(x, y, u_D) + \sum_{t'=0}^{u_D - u_A - 1} c(x, t' + u_A)$ for some x such that $(x, y) \in A$ and some $(u_A, u_D) \in \mathcal{F}(x, y, t)$.

By induction, we know that there is a feasible path $P'(s = x_1, ..., x_r = x)$ from s to x of time exactly u_A, with cost $d_b(x, u_A)$ and with zero waiting time at x. We let $u_D - u_A$ be the new waiting time at x. Since $0 \leq u_D - u_A \leq u_x$, the new path is again feasible. We extend the path with vertex y, obtaining a path P with the given waiting times and with waiting time zero at y. The time of P, with these waiting times, is exactly t, since $u_D + b(x, y, u_D) = t$, which is the arrival time at y. The cost of P with these waiting times is $d_b(x, u_A) + c(x, y, u_D) + \sum_{t'=0}^{u_D - u_A - 1} c(x, t' + u_A) = d_b(y, t)$. This prove the claim.

We now prove that $d_b(y, t)$ is the cost of a shortest feasible path from s to y with time t with waiting time zero at y. Let $P(s = x_1, ..., x_r = y)$ be a shortest feasible path from s to y of time t with waiting time zero at y. Let $w(x_i)$ be the waiting time at x_i $(i = 1, ..., r)$. So, we have $w(x_r) = 0$. Let x be the predecessor of y on this path. Let u_D be the time of the subpath P' (with the waiting times) from s to x, let $u_A = u_D - w(x)$ be the arrival time at x along P', and let $\zeta(P')$ be the cost of P'. By definition, $t = u_D + b(x, y, u_D)$. By induction, $\zeta(P') \geq d_b(x, u_A)$. By definition, the cost of P is $\zeta(P') + c(x, y, u_D) + \sum_{t'=0}^{u_D - u_A - 1} c(x, t' + u_A) \geq d_b(x, u_A) + c(x, y, u_D) + \sum_{t'=0}^{u_D - u_A - 1} c(x, t' + u_A) \geq d_b(y, t)$, where the last inequality comes from the formula on the computation of $d_b(y, t)$. This cost must be equal to $d_b(y, t)$, since P is a path of the shortest possible cost and since there exists a path that achieves $d_b(y, t)$, as we showed above. This completes the proof. $\qquad\square$

Definition 1.12 *For each arc* $(x, y) \in A$ *and each* $1 \leq t \leq T$, *define*

$$R_b(x, y, t)$$

$$= \min_{(u_A, u_D) \in \mathcal{F}(x, y, t)} \left\{ d_b(x, u_A) + c(x, y, u_D) + \sum_{t'=0}^{u_D - u_A - 1} c(x, t' + u_A) \right\},$$

and adopt the convention that $R_b(x, y, t) = \infty$ *whenever the set* $\mathcal{F}(x, y, t)$ *is empty.*

From Lemma 1.6, we have

Corollary 1.3 *For $1 \leq t \leq T$, and for each vertex y,*

$$d_b(y, t) = \min_{\{x \mid (x,y) \in A\}} R_b(x, y, t)$$

Definition 1.13 $d_b^*(x)$ *is the cost of a shortest feasible path from s to x of time at most T, subject to $w(x) = 0$ and $w(y) \leq u_y$ for any other vertex y on the path.*

From Definition 1.13, we also have $d_b^*(x) = \min_{0 \leq t \leq T} d_b(x, t)$.

In addition to the idea of sorting the values of $u + b(x, y, u)$ as discussed previously, our key idea in the algorithm to be presented below is the use of a binary heap. For every vertex x, we maintain a binary heap, denoted as $heap_x$, which contains the values of $d_b(x, u_A) - \sum_{t'=0}^{u_A - 1} c(x, t')$ for all $\max\{0, t - u_x\} \leq u_A \leq t$. Using this data structure, initialization and finding the minimum take constant time. Each insertion and each deletion take $O(\log u_x) = O(\log T)$ time. For convenience, we introduce the following notation:

Definition 1.14 $d_b^m(x, t)$ *is the minimum in the binary heap.*

We need $d_b^m(x, t)$ when evaluating $R_b(x, y, t)$. We see from Definition 1.12 that we have to solve an optimization problem of minimizing $\{d_b(x, u_A) + c(x, y, u_D) + \sum_{t'=0}^{u_D - u_A - 1} c(x, t' + u_A)\}$ subject to $(u_A, u_D) \in \mathcal{F}(x, y, t)$ to obtain $R_b(x, y, t)$. Clearly, given (x, y) and t, a value of u_D that satisfies $u_D + b(x, y, u_D) = t$ is known and, consequently, the corresponding value of $c(x, y, u_D)$ is known. Thus, solving the optimization problem reduces to solving a problem of minimizing $\{d_b(x, u_A) + \sum_{t'=0}^{u_D - u_A - 1} c(x, t' + u_A)\}$ subject to $\max\{0, u_D - u_x\} \leq u_A \leq u_D$ (recall the definition of $\mathcal{F}(x, y, t)$ in Lemma 1.6). Therefore, if $d_b^m(x, u_D)$ is known, we can obtain, for every $(x, y) \in A$ and t, $R_b(x, y, t)$, which is equal to the minimum of $d_b^m(x, u_D) + c(x, y, u_D) + \sum_{t'=0}^{u_D - 1} c(x, t')$ over all u_D satisfying $u_D + b(x, y, u_D) = t$.

We describe our algorithm for the TVSP-BWT problem as below:

Algorithm TVSP-BWT
Begin
  Initialize $d_b(s, t) = 0$ and $\forall_{x \neq s} d_b(x, t) = \infty$;
  $\forall_x$ initialize $heap_x := \{d_b(x, 0)\}$ and $d_b^m(x, 0) := d_b(x, 0)$;
  Sort all values $u + b(x, y, u)$ for $u = 0, 1, ..., T$ and for all arcs $(x, y) \in A$;
    For each $y \in V$ and each $t = 1, ..., T-1$ **do** $\mathcal{C}_y(t) := \sum_{t'=0}^{t-1} c(y, t')$;
    For $t = 1, ..., T$ **do**
      For every arc $(x, y) \in A$ **do** $R_b(x, y, t) := \infty$;
      For all $(x, y) \in A$ and all u_D such that $u_D + b(x, y, u_D) = t$ **do**

$$R_b(x, y, t) := \min\{R_b(x, y, t),$$
$$d_b^m(x, u_D) + c(x, y, u_D) + \mathcal{C}_x(u_D)\};$$

For every y **do** $d_b(y, t) := \min_{(x,y) \in A} R_b(x, y, t)$;

For every y update the $heap_y$ as follows

    Insert-heap$_y d_b(y, t) - \mathcal{C}_y(t)$;

    If $t > u_y$ **then** delete-heap$_y d_b(y, t - u_y - 1) - \mathcal{C}_y(t - u_y - 1)$;

For every y **do**

    $u_A := Minimum - heap_y$;

    $d_b^m(y, t) := d_b(y, u_A) - \mathcal{C}_y(u_A)$;

For every y **do** $d_b^*(y) := \min_{0 \le t \le T} d_b(y, t)$;

End.

Note that Algorithm TVSP-BWT computes iteratively $d_b(y, t)$ for all y at $t = 0, ..., T$. At any time t, the algorithm keeps all $d_b^m(y, u)$ for all vertices y and all $u \le t-1$. Nevertheless, for each vertex y, the algorithm maintains only one heap $heap_y$. After $d_b(y, t)$ is obtained, the new $heap_y$ at time t is obtained by deleting $d_b(y, t - u_y - 1) - \mathcal{C}_y(t - u_y - 1)$ from the heap (if $t - u_y - 1 \ge 0$) and inserting $d_b(y, t) - \mathcal{C}_y(t)$.

The following lemma is needed to show the correctness of the algorithm:

Lemma 1.7 *After the termination of Algorithm TVSP-BWT, $d_b(y, t)$ is the cost of a shortest feasible path from s to y of time exactly t and with waiting time zero at vertex y.*

Proof. We show that the formula given in Lemma 1.6 is correctly computed. Clearly, it suffices to show that $d_b^m(x, u)$, for all $0 \le u \le t$, computed by the algorithm is the minimum value of $d_b(x, u_A) - \mathcal{C}_x(u_A)$, for $u - u_x \le u_A \le u$. We use induction on t.

The argument holds for $t = 0$ because of the initialization. Now assume that the argument holds for any $0 \le u \le t - 1$, and we consider $u = t$. Note that u subject to $u + b(x, y, u) = t$ in line 8 of the algorithm must satisfy $u \le t - 1$ since $b(x, y, u)$ is positive. Thus, line 9 of the algorithm gives the correct value for $R_b(x, y, t)$ because of the assumption that $d_b^m(x, u)$ are correct for all $u \le t-1$. In addition, when $R_b(x, y, t)$ is correct, then $d_b(y, t)$ is correct according to Corollary 1.3. Consequently, Lines 12 and 13 of the algorithm generate the correct $heap_{(y)}$ at time t, and therefore $d_b^m(x, t)$ is correct. This completes the proof.   □

Lemma 1.8 *The algorithm TVSP-BWT can be implemented such that it runs in $O(T(m + n \log T))$ time.*

Proof. It is easy to see that the initialization can be done in $O(Tn)$ time.

For the sorting, we can use bucketsort, with T buckets. Since there are Tm values to be sorted, this step can then be implemented in $O(Tm)$ time.

Since the values $u + b(x, y, u)$ are now sorted, the overall time needed to update the values $R_b(x, y, t)$ is $O(Tm)$.

The two steps of inserting $d_b(y, t) - C_y(t)$ to heap$_y$ and deleting $d_b(y, t - u_y - 1) - C_y(t - u_y - 1)$ (if $t > u_y$) from heap$_y$ take $O(\log u_y) = O(\log T)$ time. Since the algorithm has to perform these two steps for all $t = 1, 2, ..., T$ and all vertex $y \in V$, it takes in total $O(Tn \log T)$ time to maintain the heaps.

The step of finding $d_b^m(y, t)$ takes $O(1)$ time. Finally, the last step of computing $d^*(y)$ for all $y \in V$ takes $O(Tn)$ time.

It follows that the overall running time of the algorithm is bounded above by $O(T(m + n \log T))$. □

Combining Lemma 1.7 with Lemma 1.8, we obtain

Theorem 1.4 *The TVSP-BWT problem with positive waiting times can be optimally solved in $O(T(m + n \log T))$ time.*

5. How to take care of the "zero" ?

In this section, we propose an aproach to handle the situation where some transit times are zero. The approach holds for all problems, TVSP-AWT, TVSP-ZWT, and TVSP-BWT. In the following, we describe it in details for the TVSP-AWT problem. The particulars of the approach for other problems can be similarly derived following the same idea. For brevity, we assume that all waiting costs are equal to zero. This assumption can be relaxed without much difficulty.

Consider a network $N(V, A, b, c)$. At the tth step of the algorithm TVSP-AWT, we first apply, as usual, the algorithm to the subnetwork $N'(V, A', b, c)$. This subnetwork N' has the same vertex set V as N, but its arc set $A' = \{(x, y) | (x, y) \in A \wedge b(x, y, t) > 0\}$. Then, after the values of $d_a(y, t)$, i.e., the costs of the shortest paths from s to each vertex y, $y \in V$, have been obtained by the algorithm TVSP-AWT, we create, for each $y \in V$, an artificial arc from s to y. Call this arc $[s, y]$. The cost of $[s, y]$ is set to $d_a(y, t)$, i.e., $c[s, y, t] = d_a(y, t)$, and the transit time on $[s, y]$ is assumed to be t. Then, we construct a new subnetwork $N''(V, A'', b, c)$. The vertex set v of the subnetwork N'' is the same as that of N. The arc set A'' consists of those arcs (x, y) for which $b(x, y, t) = 0$ and those arcs $[s, y]$ for all $y \in V \setminus \{s\}$. If there are double arcs from s to y, delete the arc from A'' which has the larger cost (or break up a tie arbitrarily if they have equal costs).

When the subnetwork N'' is created, we apply a "common" shortest path algorithm (say, Dijkstra's algorithm, see Ahuja et al (1993); Dijkstra (1959)) to N'' to find the shortest path from s to each $y \in V$. In applying such an algorithm, we ignore the transit times and the problem is thus a classical shortest path problem. For completeness, we describe the application of Dijkstra's algorithm below and refer to it as **SP**.

The algorithm maintain two sets S and S'. The set S contains vertices for which the final shortest path costs have been determined, while the set S' contains vertices for which upper bounds on the final shortest path costs are known. Initially, S contains only the source s, and the costs of the vertices in S' are set to $d_a(y, t)$. Repeatedly, select the vertex $x \notin S$, for which the distance from s is the shortest. Put x in S, and for all outgoing arcs $(x, y) \in A''$, update $d_a(y, t) := \min\{d_a(y, t), d_a(x, t) + c(x, y, t)\}$. The algorithm terminates if all vertices are in S.

We are going to show that our approach is correct in terms of finding an optimal solution at each time t for the original problem TVSP-AWT. For any vertex $y \in V$ and time $t > 0$, we can see that any path from s to y with time at most t must be one of the paths of the following type:

1. A path from s to y of time at most $t - 1$,
2. A path from s to y of time exactly t, which must pass an arc $(x, y) \in A'$ with $b(x, y, t) > 0$, or
3. A path from s to y of time exactly t, which must pass an arc $(x, y) \in A''$ with $b(x, y, t) = 0$.

In fact, for each vertex $y \in V$, our approach first uses the algorithm TVSP-AWT to determine the shortest path among those of types 1 and 2. After this is done, it creates an artificial arc $[s, y]$ to represent this shortest path. Then, it uses the procedure SP to further determine the shortest path cost of all possible paths. The shortest path can be one with only the artificial arc $[s, y]$ (in this case the algorithm TVSP-AWT had, in fact, found the optimum) or a path of type 3 (in this case, the procedure SP has found a minimal cost than that obtained by the algorithm TVSP-AWT). Since any vertex y can be reached by one of the paths of the three types, the approach has considered all possible paths and is thus optimal. Formally, we have

Lemma 1.9 *Consider the approach: At each $t = 0, 1, ..., T$, apply the algorithm TVSP-AWT to N', then apply the procedure SP to N'' to update $d_a(y, t)$ for all $y \in V$. After the t th iteration, $d_a(y, t)$ is the cost of a shortest path from s to y of time at most t.*

Proof. With induction on t, we now show that $d_a(x, t)$ is the cost of a shortest path from s to x of time at most t.

When $t = 0$, the algorithm TVSP-AW first initializes $d_a(s, 0) = 0$ and $d_a(x, 0) = \infty$ for all $x \neq s$. Then, a subnetwork N'' is created, and the procedure SP is applied to this network. Clearly, this procedure can correctly obtain, for each $x \in V$, the cost of a shortest path from s to x in the network N''. Hence, the values for $d_a(x, t)$ are correct for $t = 0$.

Now assume that the values of $d_a(x, t')$ are correct for all $x \in V$ and $t' < t$. Under this assumption, it is easy to show that the values for $d_a(x, t)$ obtained by the algorithm TVSP-AW are the costs of shortest paths of types 1 and 2. Now consider the subnetwork N'' created with artificial arcs $[s, y]$ associated with these costs. As the procedure SP is, in fact, the algorithm of Dijkstra, it can find the cost of a shortest path P from s to y in the network N'', for each $y \in V$. Moreover, the time of this path is at most t, since all arcs except those artificial arcs in N'' have zero transit times. The artificial arcs in N'' have a transit time t, but all of them originate from s and thus any path from s to y in N'' can contain at most one such arc. By the notation of our approach, $d_a(y, t)$ (updated by the procedure SP) is the cost of this shortest path P.

We now claim that $d_a(y, t)$ is also the cost of the shortest path from s to y of time at most t in the original network N. Suppose that this is not true, namely, there exists another path $\bar{P}$ from s to y, which has a cost $\zeta(\bar{P}) < d_a(y, t)$. Clearly, this cannot be a path of type 1,2, or 3; otherwise, such a path would have implied that the procedure SP, namely, Dijkstra's algorithm, is not optimal. The only possibility is that $\bar{P}$ is a path with time greater than t, which is, however, infeasible for the given t. This proves the claim, and therefore the lemma. □

For each $t = 0, 1, ..., T$, the subnetwork N' and N'' can be constructed in $O(m + n)$ time, and the procedure SP can be implemented such that it runs in $O(m + n \log n)$ time (see Ahuja et al (1993)). This is the additional running time needed to update the solutions obtained by the algorithm TVSP-AW, TVSP-ZWT, or TVSP-BWT. In summary, we have

Theorem 1.5 *The problems TVSP-AWT, TVSP-ZWT, and TVSP-BWT with non-negative transit times can be optimally solved in times $O(T(m + n \log n))$, $O(T(m + n \log n))$, and $O(T(m + n \log T + n \log n))$, respectively.*

6. Speedup to achieve an optimal time/cost trade-off

We now consider the problem in which the transit time required to traverse an arc (x, y) can be shortened, at a speedup cost. Like waiting at a vertex, speedup on an arc also enables one to catch the best timing to travel in a time-varying network. Thus, in a time-varying network, speedup may also be a better decision for the overall solution, although it incurs an extra cost locally. In particular, speedup may become necessary when the deadline T is tight. In this section we will see how the optimal solution containing speedup decisions could be computed.

Consider an arc (x, y), and let γ be the amount of time reduced from the transit time $b(x, y, t)$. Assume that the choice of γ is subject to a feasible set $\Upsilon(x, y, t)$. Corresponding to each $\gamma \in \Upsilon(x, y, t)$, there is a *speedup cost* $c_\gamma(x, y, \gamma, t)$. To be consistent with the integer assumption of the transit time $b(x, y, t)$, here we also assume that γ taken from $\Upsilon(x, y, t)$ is an integer and for any $\gamma \in \Upsilon(x, y, t)$, $b(x, y, t) - \gamma > 0$.

Let us first consider the case where waiting at a vertex is allowed without any restriction. We continue to use the notation as introduced in Section 4; for example, $d_a(y, t)$ is defined as the cost of a shortest path from s to y of time exactly t.

Lemma 1.10 $d_a(s, 0) = 0$, *and* $d_a(y, 0) = \infty$ *for all* $y \neq s$. *For* $t > 0$, *we have:*

$$d_a(y, t) = \min \Big\{ d_a(y, t - 1) + c(y, t - 1),$$

$$\min_{(x,y) \in A} \min_{(u,\gamma) \in \Gamma(x,y,t)} \{ d_a(x, u) + c(x, y, u) + c_\gamma(x, y, \gamma, u) \} \Big\}$$

where $\Gamma(x, y, t) = \{ (u, \gamma) | \gamma \in \Upsilon(x, y, u), u + b(x, y, u) - \gamma = t \}$.

The proof of Lemma 1.10 can be established by replacing $b(x, y, u)$ and $c(x, y, u)$ by $b(x, y, u) - \gamma$ and $c(x, y, u) + c_\gamma(x, y, \gamma, u)$, respectively, in the proof of Lemma 1.1.

Algorithm TVSP-AWT can be generalized as follows.

> **Algorithm TVSP-AWT-S**
> **Begin**
> Initialize $d_a(s, t) = 0$ and $\forall_{x \neq s} d_a(x, 0) := \infty$;
> Sort all values $u + b(x, y, u) - \gamma = t$ for $u = 1, 2, ..., T$, for all $\gamma \in \Upsilon(x, y, u)$ and for all arcs $(x, y) \in A$;
> > **For** $t = 1, 2, ..., T$ **do**
> > > **For** $y \in V$ **do**
> > > > **For** each $(x, y) \in A$ and each $(u, \gamma) \in \Gamma(x, y, t)$ **do**
>
> $$d_a(y, t) = \min\{ d_a(y, t - 1) + c(y, t - 1),$$
>
> $$\min_{(x,y) \in A} \min_{(u,\gamma) \in \Gamma(x,y,t)} \{ d_a(x, u) + c(x, y, u) + c_\gamma(x, y, \gamma, u) \} \}$$
>
> **Let** $d_a^*(y) := \min_{0 \leq t \leq T} d_a(y, t)$;
> **End.**

The key idea in the algorithm above is to sort, first, the values of $u + b(x, y, u) - \gamma$ for all (u, γ) and $(x, y) \in A$, where $u = 1, 2, ..., T$

and $\gamma \in \Upsilon(x, y, u)$, before the recursive relation given in Lemma 1.10 is applied to compute $d_a(y, t)$ for all $y \in V$ and $t = 1, 2, ..., T$.

Lemma 1.11 *After the termination of Algorithm TVSP-AWT-S, $d_a(y, t)$ is the cost of the shortest path from s to y of time exactly t.*

Lemma 1.11 follows from Lemma 1.10 immediately.

Lemma 1.12 *Algorithm TVSP-AWT-S can be implemented in $O(mT^2)$ time.*

Proof. To sort the values of $u + b(x, y, u) - \gamma$, we can use bucketsort, with T buckets. Since we need to check each $(x, y) \in A$, each u and each γ, there are at most mT^2 values to be sorted. Thus, this step needs $O(mT^2)$ time. Referring to the proof of Lemma 1.3, we know the running times of other steps are dominated by $O(mT^2)$, therefore, the total running time of the algorithm is bounded by $O(mT^2)$. □

Combining Lemmas 1.11 and 1.12, we have

Theorem 1.6 *The TVSP-AWT problem with speedups can be optimally solved in $O(T^2m)$ time.*

Similar to the above, we can generalize the algorithms of Section 4 to the TVSP-ZWT and TVSP-BWT problems with speedups possible for the transit times. The recursive relations for the two problems are given in the two lemmas below, respectively.

Lemma 1.13 $d_z(s, 0) = 0$, *and* $d_z(y, 0) = \infty$ *for all* $y \neq s$. *For* $t > 0$, *we have:*

$$d_z(y, t) = \min_{(x,y) \in A} \min_{(u,\gamma) \in \Gamma(x,y,t)} \{d_z(x, u) + c(x, y, u) + c_\gamma(x, y, \gamma, u)\}$$

where $\Gamma(x, y, t) = \{(u, \gamma) | \gamma \in \Upsilon(x, y, u), u + b(x, y, u) - \gamma = t\}$.

Lemma 1.14 $d_b(s, 0) = 0$, *and* $d_b(y, 0) = \infty$ *for all* $y \neq s$. *For* $t > 0$, *we have:*

$$d_b(y, t) = \min_{(x,y) \in A} \min_{(u_A, u_D, \gamma) \in \mathcal{F}^\gamma(x,y,t)} \left\{ d_b(x, u_A) \right.$$

$$+ c(x, y, u_D) + \sum_{t'=0}^{u_D - u_A - 1} c(x, t' + u_A) + c_\gamma(x, y, \gamma, u_D) \right\}$$

where $\mathcal{F}^\gamma(x, y, t) = \{(u_A, u_D, \gamma) | \gamma \in \Upsilon(x, y, u_D), u_D + b(x, y, u_D) - \gamma = t, 0 \leq u_D - u_A \leq u_x\}$.

We leave the detailed description of the algorithms to the readers.

Recall that sorting the values of $u + b(x, y, u) - \gamma$ requires $O(mT^2)$ time. Thus, from Lemmas 1.5 and 1.8, we can show:

Theorem 1.7 *The the TVSP-ZWT and TVSP-BWT problems with speedups possible for the transit times can be optimally solved by dynamic programming algorithms with time complexity of $O(mT^2)$ and $O(T(mT + n \log T))$, respectively.*

7. Additional references and comments

The main reference for this chapter is Cai, Kloks and Wong (1997). Some additional references on related studies are given below.

Cooke and Halsey (1966) consider a discrete model in which the transit time $b(x, y, t)$ varies as a function of the departure time t at vertex x, the transit cost $c(x, y, t) = b(x, y, t)$, and waiting at vertex is strictly prohibited. The function $b(x, y, t)$ is defined as a positive integer-valued function of $t \in \{t_0, t_0 + 1, ...\}$. For any vertex $x \in V$, the problem is to find the path from s to x with starting time $t = t_0$, so that the total cost (i.e., the total travel time) is minimum. They establish an optimality and develop a dynamic programming algorithm, which can solve this problem in a finite number of iterations.

Orda and Rom (1990) generalize Cooke and Halsey's model to allow for the following additional features: (a) the function of the transit time $b(x, y, t)$ is arbitrary, and (b) waiting at vertices may occur as in the following three cases:

(i) Waiting at any vertex is unrestricted; namely, unlimited waiting is allowed everywhere along the path through the network.

(ii) Waiting is forbidden; namely, waiting is disallowed everywhere along the path through the network.

(iii) Only waiting at the source vertex is allowed; namely, waiting is disallowed everywhere along the path except at the source vertex which permits unlimited waiting.

They consider an infinite time horizon (i.e., $T = \infty$). For the cases (i) and (ii) above, they provide several polynomial labeling algorithms to find the optimal solutions. They also investigate properties of the optimal paths derived and show that for case (iii) where waiting at the source vertex is arbitrary allowed, a shortest path can be found that is simple (that is, each vertex appears in the path at most once) and that can achieve a cost as cheap as the most unrestricted path (that is, the optimal solution for case (i)). Orda and Rom (1991) continue to study an infinite continuous model in which both transit time $b(x, y, t)$ and transit cost $c(x, y, t)$ are continuous functions of the departure time t at vertex x while b is strictly positive and c is nonnegative. Waiting at vertices is allowed, and a waiting cost is introduced. They set $\pi_x(t)$ as waiting cost density where t is the departure time at x, and the function of waiting cost is defined by $P_x(\alpha, t) = \int_\alpha^t \pi_x(\theta)d\theta$, where α is the arrival time at vertex x and $0 \leq \alpha \leq t \leq \infty$. They assume that, for each arc $(x, y) \in A$, there is a countable union of open and non-overlapping intervals of time, during which this arc is unavailable. Similarly, there is a countable union of open and non-overlapping intervals during which waiting is prohibited. This assumption can be regarded as an extension of the time window constraint. They point out that, the problem always has a solution no mater whether the path is finite or infinite. For the former case, an algorithm is provided that can optimally solve the problem.

Philpott and Mees (1992) examine a finite continuous shortest path model for a vehicle traveling problem, in which the transit times, parking costs and restarting costs are all time-varying but the stopping costs (waiting costs) for each unit of time is fixed. They present an algorithm and derive conditions under which the algorithm converges to an optimal solution. Psaraftis and Tsitsiklis (1993) assume that the transit costs of arcs are known functions of certain environment variables at the vertices. Each of these variables evolves according to an independent Markov process. Nachtigall (1995) study a railway model in which the transit cost $c(x, y, t)$ depends on the time t when the passenger enters the vertex x. Suppose a passenger can depart from the original station at time τ, A *transit function* $f(\tau)$ gives the earliest possible arrival time at the destination for the passenger. A label correcting method is used to calculate the desired transit function for all starting times with one path search procedure.

A type of shortest path problems that have close relation with the time-varying models we study in this chapter consider the so-called *time-window constraints*, which specify that a certain arc $(x, y) \in A$ can only be traversed within a given time period, and/or a certain vertex $x \in V$ can only be visited within a given time period.

The shortest path problem with time windows is firstly formulated as a sub-problem in the construction of school bus routes (Dosrosiers, Soumis and Desrochers (1988a)), where the tasks must be carried out according to a specified time schedule and the total cost should be minimized. Dosrosiers, Soumis and Desrochers (1988a) propose a model in which transit time $b(x, y)$ is a positive number, transit cost $c(x, y)$ is an arbitrary number, which are however all time-independent. They develop a column generation method to construct routes covering the set of tasks. At each iteration of their algorithm, the current solution is improved by inserting into the basis the least marginal cost route. This route includes a subset of tasks respecting the time window constraints and is obtained by solving the shortest path problem with time windows. Late, Desrsier and Soumis (1988b) present a labeling method to solve the problem with up to 2500 vertices and 250,000 arcs.

Sancho (1992) investigates the problem where both arcs and vertices are associated with time windows. The situation where one passes the vertex x (without waiting at x) to travel on arc (x, y) within the time window of the arc is considered feasible even though the vertex x is not visited within its time window. Sancho (1994) also considers the problem where arrival at the vertex before its time window is permitted if one is willing to wait at the vertex until the time window is opened. However, arrival at vertex after its time window is not permitted, even if no waiting is incurred at the vertex. A dynamic programming approach is developed to solve this problem.

Loachim, Gelinas, Soumis and Desrosiers (1998) propose a continuous model in which there is a vertex cost, which can be regarded as the waiting cost, at the vertex as a linear function of the service start time within the vertex time window. A dynamic programming algorithm is proposed for finding the optimal solution.

In general, a shortest path problem with time-windows can be formulated as a special case of the time-varying model we consider in this chapter, by properly setting the values for transit times and waiting constraints; see the example as illustrated in Section 1 above on the model of Desrosiers et al (1986).

Chapter 2

TIME-VARYING MINIMUM SPANNING TREES

1. Introduction

Given a network N, the *minimum spanning tree* (MST) problem is to find a connected acyclic subnetwork T that spans all the vertices of N, such that the sum of costs (or lengths) of the constituent arcs of T is minimum. The problem is further called a rooted minimum spanning tree problem, when there is a pre-specified vertex s, and the spanning tree T must have its root at s. To find the minimum spanning tree of a given network is one of the well-known problems in the field of network optimization; see Ahuja et al (1993); Graham et al (1985); Gabow et al (1986); Recski (1988).

Although the MST problem and its variants have been extensively studied in the literature, most of the works published so far treat the problem as a static one, where it is assumed that zero time is needed to travel from one vertex to another vertex, and that all attributes of the network are time-invariant. Apparently, as we indicated earlier, these assumptions are only approximations of real-world problems. In most practical situations, the network under consideration may inevitably change over time.

We will study the time-varying MST problem in this chapter. Our model considers a network where positive transit time $b(x, y, t)$ is needed to traverse an arc (x, y), at a cost $c(x, y, t)$; Moreover, both the transit time $b(x, y, t)$ and the cost $c(x, y, t)$ are time-varying, which are functions of the departure time at the vertex x. Waiting at a vertex x may be allowed, in order to catch the best timing to depart from x. Given a deadline κ and a root s, the problem is to find a rooted spanning tree to cover all vertices in the network, so that the total cost of the constituent arcs of the spanning tree is minimum, while any vertex z of the network can be reached, before the deadline κ, along a path in the spanning tree that connects s and z.

Although the static version of the MST problem is polynomially solvable (see, for example, Ahuja et al (1993)), we will show that the time-varying version is, in general, NP-complete. After laying down some basic concepts and terminologies in Section 2 below, we will focus on the MST problem over a type of arc series-parallel networks (Section 3). We will show that this problem is NP-complete in the ordinary sense (Section 3.1), and present an algorithm that can find an optimal solution in $O((m+n)m\kappa^2)$ time, where m and n are the numbers of arcs and vertices, respectively (Section 3.2). We will then consider a more general network, an undirected network containing no subgraph homomorphic to K_4 (Section 4). We will derive an algorithm that can find an exact optimal solution in $O((m+n)m\kappa^2)$ time. The general case will be studied in Section 5. Its complexity in terms of strong NP-completeness will be examined. Heuristic algorithms will be developed and their time complexity and errors will be analyzed. Finally, some additional references and remarks will be given in Section 6.

2.   Concepts and problem formulation

Let $N = (V, A, b, c)$ be a time-varying network. A vertex in N is known as the *source (root)*, denoted as s. Assume that $b(x, y, t)$ is a positive integer, $t = 0, 1, ..., \kappa$, where κ is a positive integer, representing a given time limit. By an approach similar to that of Section 5, Chapter 1, the results developed in this chapter may also be generalized to situations where the transit time b is a non-negative integer.

Recall that a dynamic path in the time-varying network N is a path P where all departure times, arrival times, and waiting times at each vertex on P are specified. We now introduce the concept of *path-induced subnetwork* as below.

Definition 2.1 *For $x_j \in V$ and $j = 1, 2, ..., J$, where $1 \leq J \leq n$, suppose $P_j(s, x_j)$ is a dynamic path from s to x_j of time at most t. Let $V(P_j)$ and $A(P_j)$ be the vertex set and the arc set of P_j, respectively, and let $V' = \bigcup_j V(P_j)$, $A' = \bigcup_j A(P_j)$. Further, let $\Gamma(P_j)$ be the set of the triples $(x, y, \tau(x))$ over P_j, where $\tau(x)$ is the departure time at vertex x on P_j, and $\Gamma = \bigcup_j \Gamma(P_j)$. Define $I_j(x) = \{[\alpha(x), \tau(x)]\}$ as the set of time intervals (the waiting times at the vertices on P_j), and $I(x) = \bigcup_j I_j(x)$. Then, $N' = (V', A', b, c)$, together with Γ and $I(x)$ ($x \in V'$), is said to be a path-induced subnetwork of N. The subnetwork N' is also said to be generalized by paths P_j, $1 \leq j \leq J$, denoted by $N' = [P_1, P_2, ..., P_J]$.*

Obviously, a single dynamic path $P(s,x)$ in N is also a path-induced subnetwork of N. On the other hand, given a path-induced subnetwork of N' of N, there must exist paths P_1, P_2, ..., P_k, such that $N' = [P_1, P_2, ..., P_k]$. The following definition gives the concept of *dynamic spanning tree.*

Definition 2.2 *Let $N' = [P_1, P_2, ..., P_J]$ be a path-induced subnetwork of N and $t \leq \kappa$. N' is said to be a dynamic spanning tree of time at most t, denoted by $T(t)$, if it satisfies the following conditions:*

(i) For each $x \in V$, there exists a dynamic path $P(s,x)$ of time at most t in N';

(ii) If x is the end vertex of path P_i in N', then x must be neither in P_i again as an intermediate vertex, nor in any other path P_j in N', where $i \neq j$ and $1 \leq i, j \leq J$.

Definition 2.3 *Let $T(t)$ be a dynamic spanning tree of time at most t, and let*

$$\zeta(T(t)) = \sum_{(x,y,\tau) \in \Gamma} c(x,y,\tau) + \sum_{x \in V} \sum_{t' \in I(x), t'=0,1,...,\kappa} c(x,t')$$

A dynamic spanning tree $T^*(t)$ is said to be a minimum spanning tree of time at most t on the time-varying network $N(V, A, b, c)$, if

$$\zeta(T^*(t)) = \min_{T(t) \in \mathcal{T}(t)} \zeta(T(t))$$

where $\mathcal{T}(t)$ is the set of all dynamic spanning trees of time at most t.

The *time-varying minimum spanning tree* (TMST) problem is to determine the minimum spanning tree $T^*(\kappa)$ for the given time-varying network N.

Example 2.1

Figure 2.1. An example of a dynamic spanning tree

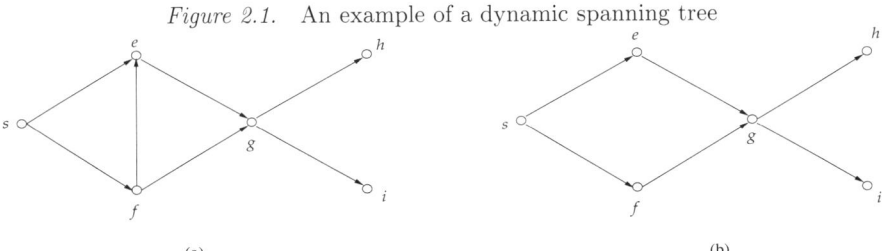

(a)   (b)

Consider a network N as shown in Figure 2.1(a), where

$$b(s, e, 0) = 1, b(s, f, 0) = 3, b(f, g, 2) = 1, b(e, g, 1) = 1,$$

$$b(f, g, 3) = 1, b(g, h, 4) = 1, b(g, i, 2) = 3,$$

$$c(s, e, 0) = 2, c(s, f, 0) = 1, c(f, g, 2) = 1, c(e, g, 1) = 1,$$

$$c(f, g, 3) = 2, c(g, h, 4) = 3, c(g, i, 2) = 2.$$

All other b and c are equal to ∞, while all $c(x, t) = 0$. Waiting at any vertex is not allowed. A dynamic spanning tree with root at s is to be found for $\kappa = 6$.

We can see that there are two dynamic paths: $P_1 = (s, e, g, i)$ with $\tau(s) = 0$, $\tau(e) = 1$ and $\tau(g) = 2$ (arrival at i at time 5); $P_2 = (s, f, g, h)$ with $\tau(s) = 0$, $\tau(f) = 3$ and $\tau(g) = 4$ (arrival at h at time 5). Let $N' = [P_1, P_2]$ be a path-induced subnetwork. From Definition 2.1, we have $V' = V$, $A' = A \backslash \{(f, e)\}$, $\Gamma = \{(s, e, 0), (s, f, 0), (e, g, 1), (f, g, 3), (g, h, 4), (g, i, 2)\}$, and $I(x) = \emptyset$ for all $x \in V$. By Definition 2.2, we can see that N' is also a dynamic spanning tree of N, since all vertices in V can be visited from the root s within time 6, and the end vertices of P_1 and P_2, i and h, only be visited once, respectively. We denote T as this tree (see Figure 2.1(b)). By Definition 2.3, the cost of T is $\zeta(T) = c(s, e, 0) + c(s, f, 0) + c(e, g, 1) + c(f, g, 3) + c(g, h, 4) + c(g, i, 2) = 11$.

The example above also shows a difference between the static spanning tree and a dynamic spanning tree as we define here. We can note that the underlying graph of T (Figure 2.1(b)) contains a cycle $C = (s, e, g, f, s)$. On the other hand, if we remove any edge from this cycle, then neither h nor i could be visited within time κ.

Remarks are given below.

Remark 2.1

(1) The model described above covers the situation where an arc (x, y) is not available during an interval $[t_1, t_2]$ (For example, the arc is not usable due to repair/maintenance work during the interval). This can be achieved by setting the value of $b(x, y, t)$ or $c(x, y, t)$ to be infinity for the interval $[t_1, t_2]$.

(2) Solomon (1986) formulates a MST model with time window constraints, where a vertex x_i ($i = 1, ..., n$) should only be visited during a given time window $[e_i, l_i]$. Early arriving at a vertex is allowed, but it must then wait at the vertex until the earliest visiting time (In other words, departure from the vertex before its earliest visiting time is prohibited).

Λ problem with time windows can be converted into the model we discuss here as follows: Suppose that the time window of a vertex y is $[e_y, l_y]$, and suppose that x_i (and z_i) are immediate preceding (and succeeding) vertices of y. Let $b(y, z_i, t) = +\infty$ for any $t < e_y$, and $b(y, z_i, t) = t_{y,z_i}$ for any $t \geq e_y$, where t_{y,z_i} is the travel time between y and z_i. Also, let $c(x_i, y, t') = d_{x_i,y}$ for any $t' + b(x_i, y, t') \leq l_y$, where $d_{x_i,y}$ is the cost of travelling from x_i to y, and $c(x_i, y, t') = +\infty$ for any $t' + b(x_i, y, t') > l_y$. Then, one can arrive at y earlier than e_y and wait at y, but he cannot depart from y before time e_y since the transit time from y to any of its succeeding vertices is infinite if the departure time is earlier than e_y. On the other hand, arriving at y later than l_y will lead to an unacceptable schedule, since the travel cost to arrive at y from any of its preceding vertices is infinite. Consequently, the optimal solution for this time-varying MST model will automatically satisfy the time window constraints.

3. Arc series-parallel networks

We will concentrate, in this section, on the TMST problem in which waiting at a vertex is arbitrary allowed, and the network has an arc series-parallel structure (see Duffin (1965); Weinberg (1971)). This type of networks have interest themselves in applications (see, e.g., application of Boolean algebra to switching circuits (Shannon 1938)). In addition, results derived on this type of networks may offer valuable insights for the study of more general problems.

A network that meets the following properties is an arc series-parallel network (ASP network):

(i) An arc (s, ρ) is an ASP network;

(ii) If G_1 and G_2 are two ASP networks and s_1, s_2 and ρ_1, ρ_2 are sources and sinks of G_1 and G_2, respectively, then G_p generated from G_1 and G_2 by merging s_1 with s_2 and ρ_1 with ρ_2 is an ASP network. Also, G_s formed from G_1 and G_2 by merging ρ_1 with s_2 is an ASP network.

In fact, any ASP network can be induced by the two properties above. Computationally, we can tell whether a given network is arc series-parallel, by using the algorithm of Valdes et al (1982), which runs in $O(m + n)$ time. An example of an arc series-parallel network is given in Figure 2.2.

In what follows, we will first study the complexity (in terms of NP-completeness) of the TMST problem over time-varying ASP networks. We will then present an exact algorithm which can find a minimum spanning tree in pseudopolynomial time. Note that when we say that

Figure 2.2.   An arc series-parallel network

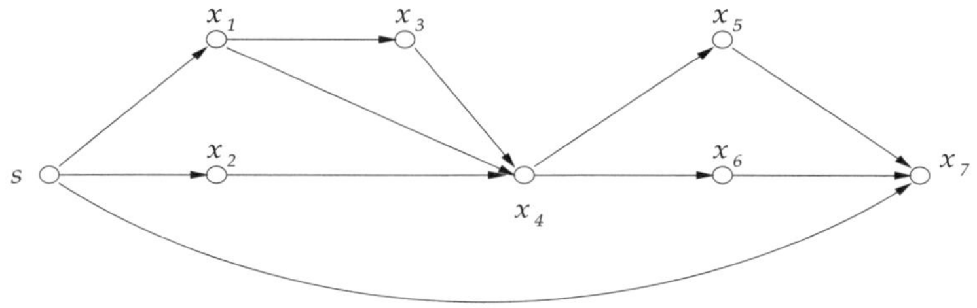

the problem we consider has an arc series-parallel structure, we always imply that the source vertex (root) of the problem is the starting vertex in the ASP network (see, e.g., Figure 2.2).

3.1   Complexity

We now show that, the TMST problem on an ASP network is NP-complete. The decision version of the problem is defined below.

TMST-SPN Given a time-varying arc series-parallel network $N(V, A, b, c)$, a time limit κ, and a threshold value K, does there exist a dynamic spanning tree $T(\kappa)$ of time at most κ, such that $\zeta(T(\kappa)) \leq K$?

We will show that the Knapsack problem (see Section 3, Chapter 1) is reducible to TMST-SPN. The following theorem establishes the NP-completeness of the TMST problem with arc series-parallel networks.

Theorem 2.1 *The TMST problem on an arc series-parallel network is NP-complete in the ordinary sense.*

Proof: For any given instance of Knapsack, we can construct a time-varying network N, with structure same as Figure 1.3, and transit times, transit costs, B, K, and time limit κ as specified in the proof of Theorem 1.1. (Note that T is the time limit used in the proof of Theorem 1.1).

We now prove that a "yes" answer to Knapsack is equivalent to a "yes" to the decision version of TMST-SPN.

If Knapsack has a set $S \subseteq Q$ such that $\sum_{i \in S} v_i \geq v^*$ and $\sum_{i \in S} w_i \leq w^*$, then we can obtain a path $P(x_0, x_{n+1})$ by the following way: starting from x_0 at time zero, for each i, if $i \in S$, then traverse arcs (x_{i-1}, x_i') and (x_i', x_i); if $i \notin S$, then traverse arc (x_{i-1}, x_i). At last, traverse arc (x_n, x_{n+1}). Let $w(x_i) = 0$ ($1 \leq i \leq n-1$) and $w(x_n) = n + w^* - \alpha(x_n)$. Notice that, since $\sum_{i \in S} w_i \leq w^*$, we have $\alpha(x_n) = \sum_{i \in S}(w_i + 1) + \sum_{i \notin S} 1 = n + \sum_{i \in S} w_i \leq n + w^*$, and $\alpha(x_{n+1}) \leq n + w^* + 1 = \kappa$, where $\alpha(x_n)$ and $\alpha(x_{n+1})$ are the arrival times of P at vertices x_n and x_{n+1}, respectively. Combining P with arcs (x_{i-1}, x_i') for each $i \notin S$, we obtain a tree, denoted as T^o. Clearly, T^o is a spanning tree of N within time duration κ. Moreover, since $\sum_{i\in S} v_i \geq v^*$, i.e., $-\sum_{i\in S} v_i \leq -v^*$, we have $\zeta(T^o) = \zeta(P) = \sum_{i\notin S} B + \sum_{i\in S}(B - v_i) = nB - \sum_{i\in S} v_i \leq nB - v^* = K$.

If TMST-SPN has a spanning tree T with the total cost not exceeding K, then there exists a path $P(x_0, x_{n+1})$ such that $\zeta(P) \leq \zeta(T) \leq K$. Let $S = \{i | (x_i', x_i) \in A(P), 1 \leq i \leq n\}$. We have $\zeta(P) = \sum_{i\in S} c(x_i', x_i, \tau(x_i')) + \sum_{i\notin S} B = \sum_{i\in S}(B - v_i) + \sum_{i\notin S} B = nB - \sum_{i\in S} v_i \leq K = nB - v^*$, which implies $\sum_{i\in S} v_i \geq v^*$. On the other hand, since T is a spanning tree of N within time κ, by the construction of the network N, we must have $\alpha(x_n) = \sum_{i\in S}(w_i + 1) + \sum_{i\notin S} 1 = n + \sum_{i\in S} w_i \leq n + w^*$, where $\alpha(x_n)$ is the arrival time of P at x_n. Therefore, $\sum_{i\in S} w_i \leq w^*$.

The analyses above show that TMST-SPN can be reduced from Knapsack and thus it is NP-complete. On the other hand, the optimal solution of TMST-SPN can be found by an algorithm in pseudopolynomial time (see Section 3.2 below) and hence it is NP-complete in the ordinary sense (cf. Garey and Johnson 1979). This completes the proof. □

3.2 A pseudo-polynomial algorithm

In an ASP network, a vertex x is called a *spreading vertex* if $d^+(x) > 1$; or a *converging vertex* if $d^-(x) > 1$, where $d^+(x)$ and $d^-(x)$ are outdegree and indegree of x, respectively. To develop our algorithm, we introduce the concept of *diamond* as follows.

Definition 2.4 *Let $f \in V$ be a spreading vertex, and $g \in V$ a converging vertex. If there are two paths from f to g containing no other spreading and converging vertices, then the subgraph induced by these two paths is called a "diamond".*

Usually, we denote $D(f, g)$ as a diamond constructed by a spreading vertex f and a converging vertex g. We also need the following definition.

Definition 2.5 *Suppose that $P(f, g)$ is a path in N which contains no other spreading and converging vertices other than f and g. For each vertex x in this path, define $d_{f,g}(x, t_s, t)$ as the cost of a shortest path from f to x so that this path can be traversed within the time duration $[t_s, t]$. If such a path does not exist, let $d_{f,g}(x, t_s, t) = \infty$.*

The following lemma gives a recursive relation to compute $d_{f,g}(x, t_s, t)$.

Lemma 2.1 *Let $P = (f, g)$ be a path which has no spreading and converging vertices other than f and g. Then, we have $d_{f,g}(f, t_s, t) = 0$ for all $0 \leq t_s \leq t \leq \kappa$, $d_{f,g}(y, t_s, t_s) = \infty$ for all $y \neq f$, and*

$$d_{f,g}(y, t_s, t) = \min\{d_{f,g}(y, t_s, t-1) + c(y, t-1),$$

$$\min_{\{u | u + b(x,y,u) = t\}} \{d_{f,g}(x, t_s, u) + c(x, y, u)\}\},$$

for $t_s < t \leq \kappa$ and $y \neq f$, where x is the predecessor of y in P.

Lemma 2.1 is a simple generalization of Lemma 1.1. For a diamond $D(f, g)$, the notation $d_{f,g}(g, t_s, t)$ may cause some confusion since there are two paths ending at g. We therefore use $d_{f,g}^{x_1}(g, t_s, t)$ and $d_{f,g}^{x_2}(g, t_s, t)$ to denote the minimum costs from f to g along paths P_1 and P_2, where x_1 and x_2 are the predecessors of g in P_1 and P_2, respectively.

Definition 2.6 *Let $D(f, g)$ be a diamond as defined in Definition 2.4 and $D' = D \backslash \{g\}$. Define $\delta_I(D, t_s, t)$ as the cost of the minimum spanning tree of D such that $\tau(f) \geq t_s$ and $\alpha(g) \leq t$. Define $\delta_E(D', t_s)$ as the cost of the minimum spanning tree of D' such that $\tau(f) \geq t_s$.*

Lemma 2.2 *Suppose D is a diamond in N and $0 \leq t_s \leq \kappa$. Then, we have*

$$\delta_I(D, t_s, t) = \min\{d_{f,g}^{x_1}(g, t_s, t) + d_{f,g}(x_2, t_s, \kappa),$$
$$d_{f,g}^{x_2}(g, t_s, t) + d_{f,g}(x_1, t_s, \kappa)\}, \quad \text{for any } t_s \leq t \leq \kappa,$$
$$\delta_E(D', t_s) = d_{f,g}(x_1, t_s, \kappa) + d_{f,g}(x_2, t_s, \kappa),$$

where x_1 and x_2 are the predecessors of g in P_1 and P_2, respectively.

Proof: Notice that $D = P_1(f, ..., x_1, g) \cup P_2(f, ..., x_2, g)$, and P_1 and P_2 have no common vertex except f and g. Thus, there are two ways to span D only, i.e., reach g by path P_1 and reach x_2 by P_2, or reach g by path P_2 and reach x_1 by P_1. Notice that if g is reached through P_1 with $\alpha(g) \leq t$, then the arrival time at vertex x_2 could be less than or equal to κ. Therefore the cost of the minimum spanning tree of D should be the minimum between these two. $\square$

The key ideas of our algorithm can be described below.

(a) If N is a path, then $d_{s,\rho}(\rho, 0, \kappa)$ is the cost of the minimum spanning tree.

(b) If N is a diamond, then $\delta_I(N, 0, \kappa)$ is the cost of the minimum spanning tree.

(c) Otherwise, choose a diamond D in N. Calculate δ_I and δ_E respectively, and make a *contracting operation* as follows:
    (i) Delete P_1 and P_2 in N except vertices f and g.
    (ii) Create an artificial vertex x' and two artificial arcs (f, x') and (x', g). Let $\delta(x', t_s, t) = \delta_E(D', t_s)$ and $\delta(g, t_s, t) = \delta_I(D, t_s, t) - \delta_E(D', t_s)$ for any $0 \leq t_s \leq t \leq \kappa$ (if both $\delta_I(D, t_s, t)$ and $\delta_E(D', t_s)$ are infinite, then $\delta(g, t_s, t) = \infty$). Then, a diamond is converted equivalently to a path $P(f, x', g)$.

(iii) It is possible that f and g are no longer a spreading vertex and a converging vertex in the new network. Suppose that the path $P(f', g')$ contains $P(f, g)$ as its subpath. Calculate $d_{f',g'}(x, t_s, t)$ for each vertex x in $P(f', g')$ according to the formula given in Lemma 2.3 below.

(d) Repeat step (c) until N becomes a path or a diamond.

Note that after a contracting operation, the new network obtained, denoted as N', contains an artificial path $P(f, g)$. Moreover, a spanning tree T' in N' should contain either an arc (f, x') or arcs (f, x') and (x', g), since $d^-(x') = d^+(x') = 1$, where $d^-(x')$ and $d^+(x')$ are in-degree and out-degree of x' in N', respectively. Therefore, we can compute the cost of T' as follows:

$$\zeta(T') = \begin{cases} \sum_{(x,y)\in A(T')\setminus(f,x')} c(x, y, \tau(x)) + \delta(x', \tau(f), \kappa), \\ \qquad\qquad\qquad\qquad\qquad \text{if } (x', g) \notin A(T') \\ \sum_{(x,y)\in A(T')\setminus\{(f,x'),(x',g)\}} c(x, y, \tau(x)) \\ \qquad + \delta(x', \tau(f), \alpha(x')) + \delta(g, \tau(f), \alpha(g)), \quad \text{otherwise} \end{cases}$$

For the case where there are multiple artificial paths, we can compute the cost of a spanning tree in a similar way. The formula given in Lemma 2.1 can be revised as follows:

Lemma 2.3 *Let $P(f, g)$ be a path, $0 \le t_s \le \kappa$, $d_{f,g}(f, t_s, t) = 0$ for all $t_s \le t \le \kappa$ and $d_{f,g}(y, t_s, t_s) = \infty$ for all $y \ne f$. For $t_s < t \le \kappa$ and $y \ne f$, we have:*

(i) If (x, y) is not an artificial arc, then

$$d_{f,g}(y, t_s, t) = \min\{d_{f,g}(y, t_s, t - 1) + c(y, t - 1),$$

$$\min_{\{u|u+b(x,y,u)=t\}} \{d_{f,g}(x, t_s, u) + c(x, y, u)\}\}$$

(ii) If (x, y) is an artificial arc, then

$$d_{f,g}(y, t_s, t) = \min\{d_{f,g}(y, t_s, t - 1) + c(y, t - 1),$$

$$\min_{t_s \le u < t} \{d_{f,g}(x, t_s, u) + \delta(y, u, t)\}\}$$

where x is the predecessor of y in P.

Let $P(f', x_1, x_2, ..., x_r, f, x', g, x_{r+1}, ..., g')$ be a path we are considering now, and $P(f, x', g)$ be a path which comes from a diamond D. For notational convenience, let us assume that $d_{s_{i+1}, \rho_{j+1}}(x', t_s, t)$ denotes the cost of the minimum spanning tree of the subnetwork generated by D'

and the path $P(f', x_1, ..., x_r, f)$, with $\tau(f') \geq t_s$, while $d_{f',g'}(g, t_s, t)$ denotes the cost of the minimum spanning tree of the subnetwork generated by D and the path $P(f', x_1, ..., x_r, f)$, with $\tau(f') \geq t_s$ and $\alpha(g) \leq t$.

We are now ready to present the following algorithm.

> **Algorithm TMST-SP**
> **Begin**
> **While** N is neither a path nor a diamond **do**
> 　　Select arbitrarily a diamond D in N;
> 　　Compute $d_{f,g}(x, t_s, t)$ for all $x \in V(D)$ and for all $0 \leq t_s \leq t \leq \kappa$;
> 　　Compute $\delta_I(D, t_s, t)$ and $\delta_E(D', t_s)$ for $0 \leq t_s \leq t \leq \kappa$;
> 　　Delete D, except the vertices f and g, from N;
> 　　Create a path $P(f, x', g)$ in N and let $\delta(x', t_s, t) = \delta_E(D', t_s)$
> and $\delta(g, t_s, t) = \delta_I(D, t_s, t) - \delta_E(D', t_s)$ for $0 \leq t_s \leq t \leq \kappa$;
> **End** *While*;
> **If** N is a path **then** $\zeta(T) = d_{s,\rho}(\rho, 0, \kappa)$;
> **If** N is a diamond **then** $\zeta(T) = \delta_I(N, 0, \kappa)$
> **End.**

Theorem 2.2 *TMST-SP can find an optimal solution for the time-varying minimum spanning tree problem with an arc series-parallel network considered in this section.*

Proof: Use induction on $|A|$. Consider the case with $m = 1$. Since there is only one arc, N is a path and the claim holds obviously. Assume that when $m < k$, the claim is true. Now we consider the case with $m = k$.

We examine the following cases:

(i) N is a path. By Lemma 2.1, the claim holds.

(ii) N is a diamond. By Lemma 2.2, the claim is also true.

(iii) N is neither a path nor a diamond. Then we select a diamond D in N, and compute $\delta_I(D, t_s, t)$ and $\delta_E(D', t_s)$ and change D to a path $P(f, x', g)$ to obtain a new network N'. Now, we prove $\zeta(T') = \zeta(T)$, where T' and T are the minimum spanning trees in N' and N, respectively.

First, we show $\zeta(T') \geq \zeta(T)$. Recall that N' differs from N, since a diamond D in N is replaced by a path $P(f, x', g)$. Because the indegree of x' equals one in N, the arc (f, x') must be in T'. Consider the arc (x', g). If $(x', g) \notin A(T')$, where $A(T')$ is the arc set of T', then we can restore the arc (f, x') to a spanning tree $T'_{D'}$ of D' with cost $\delta_E(D', t_s)$, where t_s is the departure time at f in T'. Combining $T'_{D'}$ and all other arcs in T' except (f, x'), we can obtain a spanning tree of N, denoted as T''. From Lemma 2.3, we know that the cost to reach x' from f with

$\tau(f) = t_s$ is $\delta(x', t_s, t)$. Then we have

$$\zeta(T') = \sum_{(x,y) \in A(T' \setminus (f,x'))} c(x, y, \tau(x)) + \sum_{x \in V(T' \setminus \{x'\})} \sum_{t=\alpha(x)}^{\tau(x)-1} c(x,t) + \delta(x', t_s, t)$$

$$= \sum_{(x,y) \in A(T' \setminus (f,x'))} c(x, y, \tau(x)) + \sum_{x \in V(T' \setminus \{x'\})} \sum_{t=\alpha(x)}^{\tau(x)-1} c(x,t) + \delta_E(D', t_s)$$

$$= \zeta(T'') \geq \zeta(T)$$

The last inequality holds since T is the minimum spanning tree of N. A similar analysis can be applied to the case with the arc $(x', g) \in A(T')$.

We now show $\zeta(T) \geq \zeta(T')$. For any spanning tree T^o of N, we can construct T^*, a spanning tree of N', such that $\zeta(T^o) \geq \zeta(T^*)$. Suppose that $T_{D'}^o$ is the spanning tree of D' in T^o with $t_s \leq \tau(f)$. We create T^* by replacing $T_{D'}^o$ by arc (f, x'). Let $\delta(x', t_s, t)$ denote the cost of reaching x' from f with $\tau(f) = t_s$. By the definition, we have $\delta(x', t_s, t) = \delta_E(D', t_s)$, where $\delta_E(D', t_s)$ is the cost of minimum spanning tree of D'. Therefore, we have

$$\zeta(T^o) = \sum_{(x,y) \in A(T^o \setminus T_{D'}^o)} c(x, y, \tau(x)) + \sum_{x \in V(T^o \setminus T_{D'}^o)} \sum_{t=\alpha(x)}^{\tau(x)-1} c(x,t) + \zeta(T_{D'}^o)$$

$$\geq \sum_{(x,y) \in A(T^* \setminus (f,x'))} c(x, y, \tau(x)) + \sum_{x \in V(T^o \setminus T_{D'}^o)} \sum_{t=\alpha(x)}^{\tau(x)-1} c(x,t) + \delta(x', t_s, t)$$

$$= \zeta(T^*) \geq \zeta(T').$$

Since for any T^o the inequality is true and T' is the minimum spanning tree of N', we have $\zeta(T) \geq \zeta(T')$. The proof for the case that T^o contains a spanning tree of D can be established similarly.

In summary, we have $\zeta(T') = \zeta(T)$. In other words, we can obtain the minimum spanning tree of N by finding the minimum spanning tree in N'. Notice that N' is still a time-varying arc series-parallel network with $|A(N')| \leq k - 1$ since we replace a diamond by a path and the number of arcs decreases by at least 1. By the induction, we complete the proof. $\qquad\square$

Theorem 2.3 *TMST-SP can be implemented in $O((m + n)m\kappa^2)$ time.*

Proof: The time needed for selecting a diamond in N is $O(m + n)$ (see Valdes et al (1982)). To calculate $d_{f,g}(x, t_s, t)$ for all $x \in D$ and for all $0 \leq t_s \leq t \leq \kappa$, we need $O((m + n)\kappa^2)$ time. Both the deleting and the creating operations require $O(m+n)$ time. Therefore, one iteration (within the *While* loop) needs $O((m+n)\kappa^2)$ time. Since each iteration decreases by at least one arc, we need at most m iterations. Thus, the total running time is bounded above by $O((m+n)m\kappa^2)$.   □

Let us now examine an example to illustrate Algorithm TMST-SP.

Example 2.2

Figure 2.3.   An example to illustrate Algorithm TMST-SP

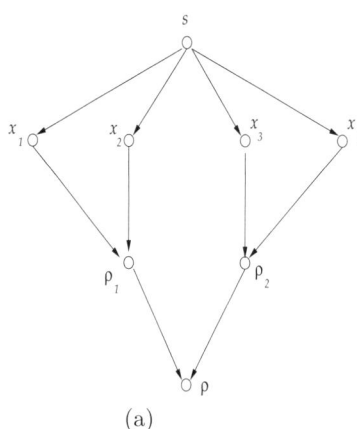

arc	t = 0 b , c	t = 1 b , c	t = 2 b , c	t = 3 b , c	t = 4 b , c
(s,x_1)	2 , 5	1 , 3	2 , 6		
(s,x_2)	1 , 1	2 , 1	2 , 2		
(s,x_3)	1 , 4	1 , 3	2 , 5		
(s,x_4)	2 , 5	2 , 3	1 , 4		
(x_1,ρ_1)			1 , 2	1 , 1	1 , 3
(x_2,ρ_1)		1 , 6	2 , 3	1 , 4	
(x_3,ρ_2)		1 , 2	2 , 1	1 , 4	
(x_4,ρ_2)			1 , 1	2 , 1	1 , 2
(ρ_1,ρ)			1 , 1	1 , 1	2 , 1
(ρ_2,ρ)			1 , 2	1 , 2	2 , 1

(a)   (b)

Consider an ASP network N as shown in Figure 2.3(a). Associated with each arc, there are two numbers, transit time $b(x,y,t)$ and cost $c(x,y,t)$, as listed in Figure 2.3(b) (a blank in the table stands for $b = c = \infty$). All waiting costs $c(x,t) = 0$. Given $\kappa = 4$, the problem is to find the time-varying minimum spanning tree T of N.

First, pick up a diamond $D_1 = (s, x_1, x_2, \rho_1)$. For each $x \in V(D_1)$ and $0 \le t_1 < t_2 \le \kappa$, calculate $d_{s,\rho_1}(x, t_1, t_2)$. For example, $d_{s,\rho_1}(s, 0, 0) = 0$, $d_{s,\rho_1}(x_2, 0, 0) = \infty$, and

$$d_{s,\rho_1}(x_2, 0, 1) = \min\{d_{s,\rho_1}(x_2, 0, 0), d_{s,\rho_1}(s, 0, 0) + c(s, x_2, 0)\}$$

$$= \min\{\infty, 0 + 1\} = 1$$

since there exists $u = 0$ which satisfies $u + b(s, x_2, u) = 1$. The values of $d_{s,\rho_1}(x, t_1, t_2)$ are shown as in Table 2.1, while $\delta_I(D_1, t_1, t_2)$ and $\delta_E(D'_1, t_1)$ are given in Table 2.2.

The network N is now converted into a network as shown in Figure 2.4(a). Then, we pick up $D_2 = (s, x_3, x_4, \rho_2)$. For each $x \in V(D_1)$ and

Table 2.1. The values of d_{s,ρ_1} for diamond D_1

$d_{s,\rho_1}(x_1,t_1,t_2)$	$t_2=1$	2	3	4	$d_{s,\rho_1}(x_2,t_1,t_2)$	$t_2=1$	2	3	4
$t_1=0$	∞	3	3	3	$t_1=0$	1	1	1	1
1		3	3	3	1		∞	1	1
2			∞	6	2			∞	2
3				∞	3				∞

$d_{s,\rho_1}^{x_1}(\rho_1,t_1,t_2)$	$t_2=1$	2	3	4	$d_{s,\rho_1}^{x_2}(\rho_1,t_1,t_2)$	$t_2=1$	2	3	4
$t_1=0$	∞	∞	5	4	$t_1=0$	∞	7	7	4
1		∞	5	4	1		∞	∞	5
2			∞	∞	2			∞	∞
3				∞	3				∞

Table 2.2. The values of δ_I and δ_E for diamond D_1

$\delta_I(D_1,t_1,t_2)$	$t_2=1$	2	3	4	t_1	$\delta_E(D_1',t_1)$
$t_1=0$	∞	10	6	5	0	4
1		∞	6	5	1	4
2			∞	∞	2	8
3				∞	3	∞

$0 \le t_1 < t_2 \le \kappa$, calculate $d_{s,\rho_2}(x,t_1,t_2)$, see Table 2.3. The values of $\delta_I(D_2,t_1,t_2)$ and $\delta_E(D_2',t_1)$ are listed in Table 2.4.

A new network is obtained as shown in Figure 2.4(b).

Finally, as the network is a diamond now, we can calculate $\delta_I(D,t_1,t_2)$ for $0 \le t_1 < t_2 \le \kappa$. The results are listed in Table 2.5.

Since $\delta_I(D,0,4) = 14$, we have $\zeta(T) = 14$. The minimum spanning trees T is shown as in Figure 2.5(a). The number in a box associated with the arc is the departure time and the number without a box is the transit cost. Figure 2.5(b) shows a minimum spanning tree of N with $\kappa = 3$.

Table 2.3. The values of d_{s,ρ_2} for diamond D_2

$d_{s,\rho_2}(x_3,t_1,t_2)$	$t_2 = 1$	2	3	4	$d_{s,\rho_2}(x_4,t_1,t_2)$	$t_2 = 1$	2	3	4
$t_1 = 0$	4	3	3	3	$t_1 = 0$	∞	5	3	3
1		3	3	3	1		∞	3	3
2			∞	5	2			4	4
3				∞	3				∞
$d_{s,\rho_2}^{x_3}(\rho_2,t_1,t_2)$	$t_2 = 1$	2	3	4	$d_{s,\rho_2}^{x_4}(\rho_2,t_1,t_2)$	$t_2 = 1$	2	3	4
$t_1 = 0$	∞	6	6	4	$t_1 = 0$	∞	∞	6	6
1		∞	∞	4	1		∞	∞	∞
2			∞	∞	2			∞	∞
3				∞	3				∞

Table 2.4. The values of δ_I and δ_E for diamond D_2

$\delta_I(D_2,t_1,t_2)$	$t_2 = 1$	2	3	4	t_1	$\delta_E(D_2',t_1)$
$t_1 = 0$	∞	9	9	7	0	6
1		∞	∞	7	1	6
2			∞	∞	2	9
3				∞	3	10

Table 2.5. The values of δ_I for diamond D

$\delta_I(D,t_1,t_2)$	$t_2 = 1$	2	3	4
$t_1 = 0$	∞	∞	16	14
1		∞	∞	14
2			∞	∞
3				∞

Figure 2.4. An example to illustrate Algorithm TMST-SP (continued)

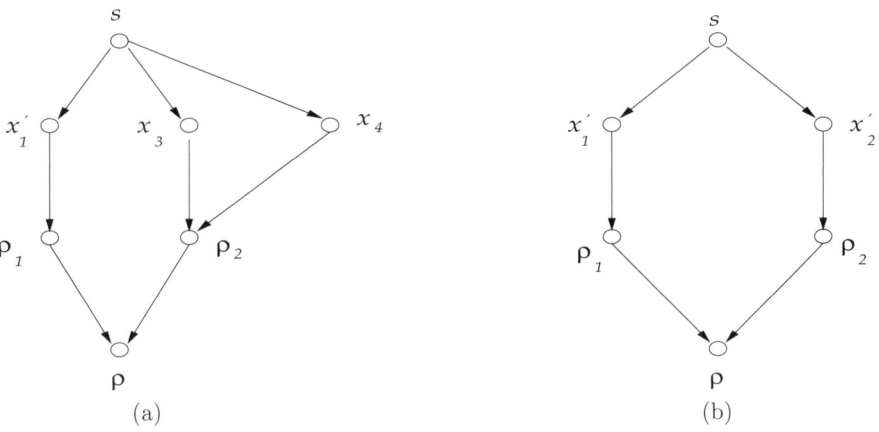

Figure 2.5. The optimal solutions

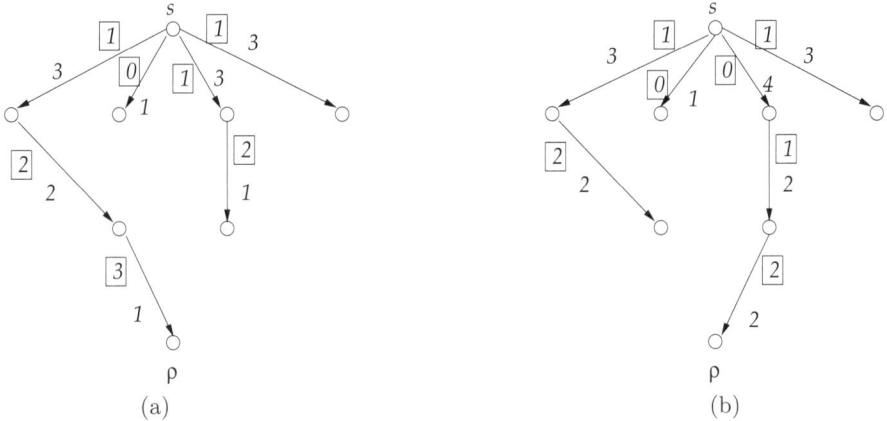

4. Networks containing no subgraph homomorphic to K_4

In this section, we will generalize the model of Section 3.3 to such networks whose underlying graphs contain no subgraphs homomorphic to K_4 (a complete graph with four vertices).

4.1 Properties and complexity

Liu and Geldmacher (1976) have shown that, any graph with no subgraph homomorphic to K_4 can be recursively transformed, by applying four transformation rules (see Definition 2.7 below), to a single vertex. They have further devised a linear time algorithm that can decide whether a graph has a subgraph homomorphic to K_4 (Liu and Geldmacher 1980).

Definition 2.7 *Let G' be the resultant graph after applying four transformation rules T_1, T_2, T_3 and T_4 to a graph G until none of the rules can be further applied, where*

T_1: *Replace a loop vv with a vertex v.*

T_2: *Replace a dangling edge uv with a vertex u.*

T_3: *Replace a pair of series edges uv and vw with an edge uw.*

T_4: *Replace a pair of parallel edges uv and uv with an edge uv.*

If G' consists of only one single vertex, then we say G is reducible. Otherwise, we say G is nonreducible.

Note that in the definition above we follow Liu and Geldmacher (1976) to use the terminology "reducible", which is different from the terminology "reducible" used in the NP-completeness analysis (see Section 3.2).

The following two properties are established in Liu and Geldmacher (1976).

Property 2.1 *If T_1, T_2, T_3 and T_4 are applied to a graph until no longer possible, then a unique graph results, independent of the sequence of application of T_1, T_2, T_3 and T_4.*

Property 2.2 *A graph G is nonreducible if and only if it contains a subgraph homomorphic to K_4.*

Corresponding to the concept of reducible graph, we define the terminology of "reducible network" as follows.

Definition 2.8 *A is a **reducible network** if its underlying graph contains no subgraph homomorphic to K_4.*

Note that an edge series-parallel graph (the underlying graph of an arc series-parallel network) is reducible since it contains no subgraph homomorphic to K_4. On the other hand, we do have other networks, whose underlying graph are not edge series-parallel, but which are reducible. Figure 2.6 below is an example.

Example 2.3

An ASP network is a special case of the reducible network. Thus, from Theorem 2.1 and Theorems 2.5 and 2.6 (See below), we have the following Theorem.

Theorem 2.4 *The time-varying spanning tree problem on reducible networks is NP-complete in the ordinary sense.*

Figure 2.6. A reducible graph which is not edge series-parallel

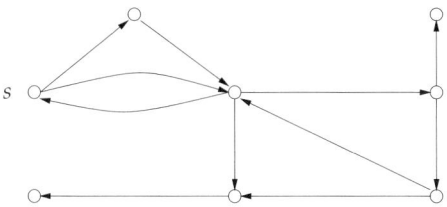

Figure 2.7. Two special cases of reducible networks

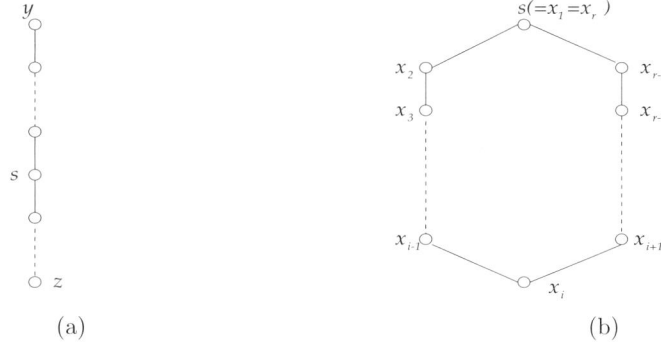

(a)   (b)

4.2 An exact algorithm

Recall that we consider time-varying networks with no parallel arcs, that is, there do not exist two arcs of the same direction between two vertices. It is possible to have, however, two arcs of opposite directions between two vertices. To simplify the presentation in figures, in this section we will use a link to indicate a single arc or a pair of opposite arcs between two vertices.

Let us first examine the following two special cases.

Case I. The network N under consideration is shown in Figure 2.7(a), where s is the source vertex. By Definition 2.5, $d_{s,x}(x, t_s, t)$ is the cost of a shortest path from s to x within the time duration $[t_s, t]$ (Note that if such a path does not exist, then $d_{s,x}(x, t_s, t) = \infty$). Then, we can easily see that

$$\zeta(T(\kappa)) = d_{s,y}(y, 0, \kappa) + d_{s,z}(z, 0, \kappa) \tag{2.1}$$

where $\zeta(T(\kappa))$ is the minimum cost of the spanning tree of N within the time limit κ.

Case II. The network N under consideration is shown in Figure 2.7(b). In this case we have

$$\zeta(T(\kappa)) = \min_{1 \leq i \leq r-1} \{d_{s,x_i}(x_i, 0, \kappa) + d_{s,x_{i+1}}(x_{i+1}, 0, \kappa)\} \qquad (2.2)$$

where $d_{s,x_i}(x_i, 0, \kappa)$ is calculated along the path $P(x_1, x_2, ..., x_i)$ while $d_{s,x_{i+1}}(x_{i+1}, 0, \kappa)$ is calculated along the path $P(x_r, x_{r-1}, ..., x_{i+1})$. Again, note that by default we define $d_{s,x}(x, t_s, t) = \infty$ if no path exists from s to x.

A vertex $v \in V$ is said to be a *conjunction vertex*, if it has more than two adjacent vertices in N. A path $P(x, y)$ (or a cycle $C(x = x_1, ..., x_r = x)$) with one conjunction vertex x and $s \notin V(P)\backslash\{x\}$ (or $s \notin V(C)\backslash\{x\}$) is called a *dangling path (or a dangling cycle)* (see Figure 2.8). A cycle with two conjunction vertices is called a *diamond*. Note that this definition is a generalization of Definition 2.4.

Figure 2.8. l and e are conjunction vertices, while $P(a, e)$ is a dangling path, cycle $C(e, g, h)$ is a dangling cycle, and $D(P(s, w, l), P(s, f, l))$ is a diamond.

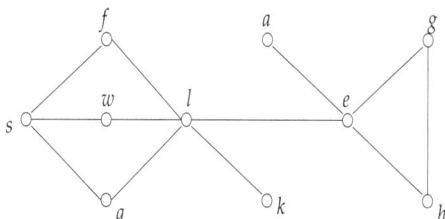

Both two structures, path and diamond, play the key roles in the algorithm we are to present below. Notice that, if a path contains two conjunction vertices as its two ends and s is not in it, the flow can flow in or out at each conjunction vertex of the path. Figure 2.9 shows these situations.

Figure 2.9. Directions of flow into or out from a path, where y and z are two conjunction vertices

Since for two vertices x and y, $c(x, y, t)$ may not equal $c(y, x, t)$, we need to calculate the cost of a path in terms of three different spanning ways above. This leads to the following definition.

Definition 2.9 *Suppose that $P(y = x_1, x_2, ..., x_r = z)$ is a path in N, where y and z are two conjunction vertices, and $s \notin V(P) \backslash \{y, z\}$. Define*
(1) $d_{y^I, z^O}(t_y, t_z)$ as the cost of the minimum spanning tree of P, such that y is the flow-in vertex, z is the flow-out vertex, $\tau(y) \geq t_y$ and $\alpha(z) \leq t_z$;
(2) $d_{y^O, z^I}(t_y, t_z)$ as the cost of the minimum spanning tree of P, where y is the flow-out vertex, z is the flow-in vertex, $\alpha(y) \leq t_y$ and $\tau(z) \geq t_z$;
(3) $d_{y^I, z^I}(t_y, t_z)$ as the minimum forest of P, where both y and z are flow-in vertices, $\tau(y) \geq t_y$ and $\tau(z) \geq t_z$.
If such trees do not exist, let the relevant cost be ∞.

Recall Definition 2.5 on $d_{f,g}(x, t_s, t)$. The following lemma gives a method to calculate $d_{y^I, z^O}(t_y, t_z)$, $d_{y^O, z^I}(t_y, t_z)$, and $d_{y^I, z^I}(t_y, t_z)$.

Lemma 2.4 *Suppose that $P(y = x_1, x_2, ..., x_r = z)$ is a path in N, where y and z are two conjunction vertices, and $s \notin V(P) \backslash \{y, z\}$. Then, we have*

$$d_{y^I, z^O}(t_y, t_z) = d_{y,z}(z, t_y, t_z), \quad 0 \leq t_y \leq t_z \leq \kappa$$

$$d_{y^O, z^I}(t_y, t_z) = d_{z,y}(y, t_z, t_y), \quad 0 \leq t_z \leq t_y \leq \kappa$$

$$d_{y^I, z^I}(t_y, t_z) = \min_{1 \leq i \leq r-1} \{d_{y,z}(x_i, t_y, \kappa) + d_{z,y}(x_{i+1}, t_z, \kappa)\}, \quad 0 \leq t_y, t_z \leq \kappa.$$

Proof: Straightforward. □

By Lemma 2.4, formulae (2.1) and (2.2) can be rewritten as (2.3) and (2.4) respectively.

$$\zeta(T^*) = d_{s^I, y^O}(0, \kappa) + d_{s^I, z^O}(0, \kappa), \tag{2.3}$$

$$\zeta(T^*) = \min_{1 \leq i \leq r-1} \{d_{s^I, x_i^O}(0, \kappa) + d_{s^I, x_{i+1}^O}(0, \kappa)\} \tag{2.4}$$

Similarly, if a diamond does not contain the source vertex, the flow can also flow in or out at each conjunction vertex of a diamond. Figure 2.10 shows these situations.

Definition 2.10 *Suppose that D is a diamond in N, where y and z are two conjunction vertices, and $s \notin V(D) \backslash \{y, z\}$. Define $\delta_{y^I, z^O}(t_y, t_z)$ as the cost of the minimum spanning tree of D such that y is the flow-in vertex, z is the flow-out vertex, $\tau(y) \geq t_y$ and $\alpha(z) \leq t_z$. Define $\delta_{y^O, z^I}(t_y, t_z)$ and $\delta_{y^I, z^I}(t_y, t_z)$ in a similar way. If such spanning trees do not exist, let the cost be ∞.*

Figure 2.10. Directions of flow into or out from a diamond, where y and z are two conjunction vertices

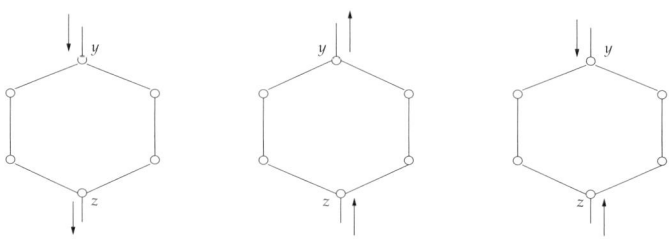

Lemma 2.5 *Suppose that* $D = (P_1(y, z), P_2(y, z))$ *is a diamond which consists of two paths* P_1 *and* P_2*, with* y *and* z *being the two conjunction vertices, and* $s \notin V(D) \backslash \{y, z\}$*. Then, we have*

$$\delta_{y^I, z^O}(t_y, t_z) = \min\{d^1_{y^I, z^O}(t_y, t_z) + d^2_{y^I, z^I}(t_y, t_z), d^2_{y^I, z^O}(t_y, t_z)$$

$$+ d^1_{y^I, z^I}(t_y, t_z)\}, \qquad\qquad 0 \le t_y \le t_z \le \kappa$$

$$\delta_{y^O, z^I}(t_y, t_z) = \min\{d^1_{y^O, z^I}(t_y, t_z) + d^2_{y^I, z^I}(t_y, t_z), d^2_{y^O, z^I}(t_y, t_z)$$

$$+ d^1_{y^I, z^I}(t_y, t_z)\}, \qquad\qquad 0 \le t_z \le t_y \le \kappa$$

$$\delta_{y^I, z^I}(t_y, t_z) = d^1_{y^I, z^I}(t_y, t_z) + d^2_{y^I, z^I}(t_y, t_z), \qquad 0 \le t_y, t_z \le \kappa$$

where d^k *denotes the cost for* P_k *(*$k = 1, 2$*).*

Proof: Straightforward. □

The basic idea of our algorithm is to carry out two operations, one transferring a dangling path or a dangling cycle to a vertex, while the other transferring a diamond to a path.

Operation I Consider a dangling path (or a dangling cycle) of N. Before give the operation, we introduce the following definition first.

Definition 2.11 *Suppose* $P(x, ..., y)$ *be a dangling path in* N *and* x *is a conjunction vertex. Define*

$$\eta(x, t) = d_{x^I, y^O}(t, \kappa)$$

where $0 \le t \le \kappa$*. In case* $y = x$*,* P *becomes a dangling cycle* $C(x, ..., x)$*. Then, define*

$$\eta(x, t) = d_{x^I, x^I}(t, t).$$

Actually, $\eta(x, t)$ is the cost of the shortest path of P or the cost of the minimum spanning tree of C with departure time t at vertex x. The operation is to remove the path P from N (except vertex x), and attach $\eta(x,t)$ $(0 \leq t \leq \kappa)$ to vertex x in N'. For completeness, we let $\eta(z,t) = 0$ $(0 \leq t \leq \kappa)$ for any $z \in V(N')$ if there does not exist a dangling path or a dangling cycle at z. Furthermore, If there are more than one dangling paths or dangling cycles at vertex x, let $\eta(x,t)$ be the summation of the costs.

Operation II Consider a diamond $D(P_1(e,...,g), P_2(e,...,g))$, where e and g are two conjunction vertices. Now, we remove D in N except vertices e and g, and add an artificial path $P(e,x',g)$ into N to obtain a new network N', where x' is an artificial vertex, ex' and $x'g$ are two artificial edges (see Figure 2.11). Moreover, attach $\delta_{e^I,g^O}(t_e, t_g)$, $\delta_{e^I,g^O}(t_e, t_g)$, and $\delta_{e^I,g^O}(t_e, t_g)$ to vertex x'.

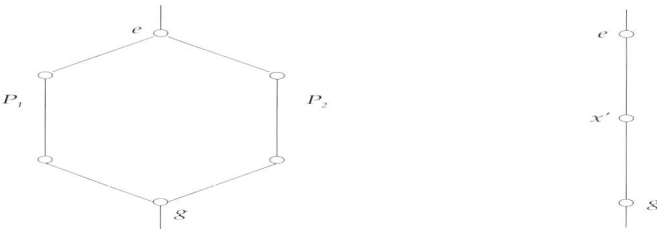

Figure 2.11. A diamond is changed into a path

Notice that, the network N is reducible, therefore after performing these two operations until no longer possible, the resultant network is either a path or a cycle as described in case I and II.

After completing operations, N' contains artificial vertices, artificial edges and extra variables $\eta(x,t)$. Therefore, we need to define the cost of a spanning tree of N' as below:

Definition 2.12 *Suppose N' is a network obtained after the operations, and T' be a spanning tree of N'. Define*

$$\zeta(T') = \sum_{xy \in E(T'), x,y \notin \mathcal{A}(N')} c(x,y,\tau(x)) + \sum_{x \in V(T')} \eta(x, \alpha(x)) + \sum_{x \in \mathcal{A}(N')} \Delta(x)$$

where

$$\Delta(x) = \begin{cases} \delta_{e^I,g^I}(\alpha(e), \alpha(g)), & \text{if } (e,x') \in E(T'), (x',g) \notin E(T'), \\ & \text{or } (x',g) \in E(T'), (e,x') \notin E(T') \\ \delta_{e^I,g^O}(\alpha(e), \tau(g)), & \text{if } (e,x'), (x',g) \in E(T'), \alpha(e) \leq \tau(g) \\ \delta_{e^O,g^I}(\tau(e), \alpha(g)), & \text{if } (e,x'), (x',g) \in E(T'), \tau(e) > \alpha(g) \end{cases}$$

e and g are conjunction vertices of D, and $\mathcal{A}(N')$ is the artificial vertex set of N'.

For the case where there are multiple artificial vertices, we can define the cost of a spanning tree in a similar way. Definition 2.12 is the generalization of Definition 2.3. One can see that if N' has no artificial paths and all $\eta = 0$, then Definition 2.12 becomes Definition 2.3.

Furthermore, the formula for calculating $d_{e,g}(x, t_e, t_g)$ and $d_{g,e}(x, t_g, t_e)$ should be revised as follows:

Lemma 2.6 *Let $P(e, ..., x, y, z, ..., g)$ be a path of N where e and g are two conjunction vertices. Then, $d_{e,g}(e, t_e, t) = 0$ for all $0 \leq t_e \leq t \leq \kappa$, $d_{e,g}(y, t_e, t_e) = \infty$ for all $y \neq e$, and, for $0 \leq t_e < t \leq \kappa$ and $y \neq e$, we have*

(i) if xy is not an artificial edge, then

$$d_{e,g}(y, t_e, t) = \min\{d_{e,g}(y, t_e, t - 1),$$

$$\eta(y, t) + \min_{\{u|u+b(x,y,u)=t\}}\{d_{e,g}(x, t_e, u) + c(x, y, u)\}\},$$

(ii) if xy is an artificial edge, and
    (a) if y is an artificial vertex, then

$$d_{e,g}(y, t_e, t) = \min\{d_{e,g}(y, t_e, t-1), \min_{t_e<u<t}\{d_{e,g}(x, t_e, u) + \delta_{x^I, z^O}(u, t)\}\},$$

    (b) otherwise, if y is not an artificial vertex, then

$$d_{e,g}(y, t_e, t) = \eta(y, t) + \min\{d_{e,g}(y, t_e, t - 1), d_{e,g}(x, t_e, t)\},$$

Similarly, we can compute $d_{g,e}(x, t_g, t_e)$.

Also, Lemma 2.4 should be revised as follows:

Lemma 2.7 *Suppose that $P(y = x_1, x_2, ..., x_r = z)$ is a path in N, where y and z are two conjunction vertices, and $s \notin V(P) \backslash \{y, z\}$. Then, we have*

$$d_{y^I, z^O}(t_y, t_z) = d_{y,z}(z, t_y, t_z), \qquad 0 \leq t_y \leq t_z \leq \kappa$$
$$d_{y^O, z^I}(t_y, t_z) = d_{z,y}(y, t_z, t_y), \qquad 0 \leq t_z \leq t_y \leq \kappa$$
$$d_{y^I, z^I}(t_y, t_z) = \min_{1 \leq i \leq r-1} \xi_i, \qquad 0 \leq t_y, t_z \leq \kappa$$

where

$$\xi_i = \begin{cases} d_{y,z}(x_i, t_y, \kappa) + d_{z,y}(x_{i+1}, t_z, \kappa), \\ \qquad \text{both } x_i \text{ and } x_{i+1} \text{ are not artificial vertices} \\ \min_{t_y \leq u_1 \leq \kappa, t_z \leq u_2 \leq \kappa}\{d_{y,z}(x_{i-1}, t_y, u_1) + \delta_{x_{i-1}^I, x_{i+1}^I}(u_1, \kappa) \\ \qquad + d_{z,y}(x_{i+1}, t_z, \kappa)\}, \qquad x_i \text{ is an artificial vertex} \\ \min_{t_y \leq u_1 \leq \kappa, t_z \leq u_2 \leq \kappa}\{d_{y,z}(x_i, t_y, \kappa) + \delta_{x_i^I, x_{i+2}^I}(\kappa, u_2) \\ \qquad + d_{z,y}(x_{i+2}, t_z, u_2)\} \qquad x_{i+1} \text{ is an artificial vertex} \end{cases}$$

By Lemma 2.6 and Lemma 2.7, formulae (2.3) and (2.4) now should be rewritten as (2.5) and (2.6) respectively.

$$\zeta(T^*) = d_{s^I,y^O}(0,\kappa) + d_{s^I,z^O}(0,\kappa) + \eta(s,0), \qquad (2.5)$$

$$\zeta(T^*) = d_{s^I,s^I}(0,0) + \eta(s,0). \qquad (2.6)$$

The basic steps of our algorithm are as follows:

(a) If N is a path $P(y, ..., s, ..., z)$, then $\zeta(T(\kappa))$ is calculated by formula (2.5) (Assume the initial value of $\eta(x,t)$ is set to zero for each $x \in V$ and $t = 0, 1, ..., \kappa$).

(b) If N is a cycle $C(s = x_1, x_2, ..., x_r = s)$, then $\zeta(T(\kappa))$ is calculated by formula (2.6).

(c) If N contains a dangling path $P(x, y)$, where x is a conjunction vertex, then let $\eta(x,t) = d_{x^I,y^O}(t,\kappa) + \eta(x,t)$ for $t = 0, 1, ..., \kappa$, and do Operation I.

(d) If N contains a dangling cycle $C(x = x_1, x_2, ..., x_r = x)$, where x is a conjunction vertex, then for $t = 0, 1, ..., \kappa$, let $\eta(x,t) = d_{s^I,s^I}(t,t) + \eta(s,t) + \eta(x,t)$, and do Operation I.

(e) Otherwise, choose a diamond $D(P_1(y, z), P_2(y, z))$ in N (if there are more than one diamond, choose one arbitrarily), where y and z are two conjunction vertices of D. According to Lemma 2.5, calculate $\delta_{y^I,z^O}(t_y, t_z)$ for $0 \le t_y \le t_z \le \kappa$, $\delta_{y^O,z^I}(t_y, t_z)$ for $0 \le t_z \le t_y \le \kappa$, and $\delta_{y^I,z^I}(t_y, t_z)$ for $0 \le t_y, t_z \le \kappa$. Perform Operation II.

(f) Still denote the new network as N. Repeat step (c) to (e) until N becomes a path or a cycle.

Now, we are ready to present our algorithm.

Algorithm TMST-RN
Begin
Set $\eta(x,t) = 0$ for each $x \in V(N)$ and for each $0 \le t \le \kappa$;
While N is neither a path nor a cycle with s **do**
Repeat doing Operation I to delete the dangling paths and dangling cycles;
Select arbitrarily a diamond $D(y, z)$ in N;
Calculate $\delta_{y^I,z^O}(t_y, t_z)$, $\delta_{y^O,z^I}(t_y, t_z)$, and $\delta_{y^I,z^I}(t_y, t_z)$ for D and for all $0 \le t_y, t_z \le \kappa$;
Do Operation II;

End *While*;
    If N is a path $P(x, ..., s, ..., y)$ **then** $\zeta(T) = d_{s^I,y^O}(0,\kappa) + d_{s^I,z^O}(0,\kappa) + \eta(s,0)$;
    If N is a cycle $C(s = x_1, x_2, ..., x_r = s)$ **then** $\zeta(T) = d_{s^I,s^I}(0,0) + \eta(s,0)$;
    End.

Theorem 2.5 *TMST-RN can optimally solve the time-varying minimum spanning tree problem on a reducible network.*

Proof: We prove that, when the algorithm is terminated, $\zeta(T)$ is the cost of the minimum spanning tree of N. Use induction on $m = |E(N)|$. Consider the case with $m = 1$. Since there is only one arc, N is a path and the claim holds obviously. Assume that when $m < k$, the claim is true. Now we consider the case with $m = k$.

We examine the following cases:

(1) N is a path or a cycle. By formulae (2.5) and (2.6), we know that $\zeta(T)$ is the cost of the minimum spanning tree of N.

(2) N is neither a path nor a cycle. Then consider the following cases:

(i) N has a dangling path $P(x, y)$, where x is a conjunction vertex. By the algorithm, we delete P (except vertex x) from N and obtain a new network N'. In what follows, we first prove that, for each spanning tree T' of N', it can be extended to a spanning tree T^o of N such that $\zeta(T') = \zeta(T^o)$. Next, we show that for each spanning tree T of N, there is a spanning tree T' of N', such that $\zeta(T') \leq \zeta(T)$. That is to say, finding the minimum spanning tree of N can be equivalently converted to finding the minimum spanning tree of N'. Since $m' = |E(N')| < k$, by the induction, $\zeta(T')$ obtained by the algorithm is the cost of minimum spanning tree of N', therefore, it is also the cost of the minimum spanning tree of N.

Suppose T' is a spanning tree of N'. By the definition, we have $\zeta(T') < \infty$. By the definition of $\zeta(T')$, we know that $\sum_{x' \in V(T')} \eta(x', \alpha(x')) < \infty$, or, $\eta(x, \alpha(x)) < \infty$. By the definition of η, we know that there is a spanning tree T'' of $P(x, y)$ with the cost $\eta(x, \alpha(x))$. Thus, we can extend T' by adding T'' to obtain a dynamic spanning tree of N.

Now, we show that $\zeta(T') \leq \zeta(T)$. Notice that P is a dangling path in N, that is to say, all vertices y in P can only be visited from x. Suppose that t^o be the departure time at x in T. We create T' by cutting path P in T except vertex x, and let $\eta(x, t^o)$ be the minimum spanning tree of path P with departure time t^o at vertex x. Then, we have $\eta(x, t^o) \leq \sum_{uv \in E(P)} c(u, v, \tau(u))$. By the definition, we have

$$\zeta(T') = \sum_{x'y' \in E(T')} c(x', y', \tau(x')) + \sum_{x \in V(T')} \eta(x', \alpha(x'))$$

$$= \sum_{x'y' \in E(T')} c(x', y', \tau(x')) + \sum_{x \in V(T'), x' \neq x} \eta(x', \alpha(x')) + \eta(x, t^o)$$

$$\leq \sum_{x'y' \in E(T')} c(x', y', \tau(x')) + \sum_{x \in V(T'), x' \neq x} \eta(x', \alpha(x')) + \sum_{uv \in E(P)} c(u, v, \tau(u))$$

$$= \zeta(T)$$

Therefore, the claim is proved.

(ii) N has a dangling cycle. The analysis is similar to (i).

(iii) N has neither dangling paths nor dangling cycles. Since N is reducible, there must exist a diamond with two conjunction vertices only (Otherwise, N will contain a subgraph homomorphic to K_4 since the transformation T_4 can not be applied to N).

Select a diamond D in N with two conjunction vertices, say e and g. We compute $\delta_{e^I, g^O}(t_1, t_2)$, $\delta_{e^O, g^I}(t_1, t_2)$, and $\delta_{e^I, g^I}(t_1, t_2)$. Change D to a path $P(e, x', g)$ to obtain a new network N'. Similarly, we first prove that for any spanning tree T' of N", it can be extended to a spanning tree T of N.

Notice that in N', the degree of x' is 2, therefore edge ex' (or $x'g$) must be in T'. Suppose ex' in T' and $x'g \notin T'$. We restore edge ex' and vertex g to a spanning forest F of the diamond D, and add F to T' to obtain a spanning tree T of N. A similar analysis can be applied to the case with $ex', x'g \in E(T')$ or $x'g \in E(T')$ only.

We now show that for any spanning tree T of N, there is a spanning tree T' of N', such that $\zeta(T') \leq \zeta(T)$. Suppose that F is the spanning forest of D in T with $t_e = \tau(e)$, and $t_g = \tau(g)$. We create T' by replacing F by edge ex'. Let $\delta_{e^I, g^I}(t_e, t_g)$ denote the cost of minimum spanning forest of D. Therefore, we have

$$\zeta(T) = \sum_{xy \in E(T) \setminus E(F)} c(x, y, \tau(x)) + \zeta(F)$$

$$\geq \sum_{xy \in E(T') \setminus ex'} c(x, y, \tau(x)) + \delta_{e^I, g^I}(t_e, t_g) = \zeta(T').$$

The proof for the case that T contains a spanning tree of D can be established similarly.

Notice that N' is still a time-varying reducible network with $|E(N')| < k$, since we replace a diamond by a path or replace a path (a cycle) to a vertex and the number of edges decreases by at least 1. By the induction, we complete the proof.

$\square$

Theorem 2.6 *TMST-RN can be implemented in* $O((m+n)m\kappa^2)$ *time.*

Proof: Setting $\eta(x,t)$ for each vertex $x \in V$ and for each time $\leq t \leq \kappa$ needs $O(n\kappa)$ time. In the while-loop, to find a dangling path or a dangling cycle, we can use depth-first traversal (see Gilberg et al (2001)). This step can be done in $O(m)$ time. The time needed for selecting a diamond in N is $O(m+n)$ (see Valdes et al (1982)). To calculate $\delta_{y^I,z^O}(t_1,t_2)$, $\delta_{y^O,z^I}(t_1,t_2)$, and $\delta_{y^I,z^I}(t_1,t_2)$ for all $0 \leq t_1 \leq t_2 \leq \kappa$, we need $O((m+n)\kappa^2)$ time. Replacing a diamond by an artificial path requires $O(m)$ time. Therefore, one iteration (within the *While* loop) needs $O((m+n)\kappa^2)$ time. Since each iteration decreases at least one edge, we need at most m iterations. Thus, the total running time is bounded above by $O((m+n)m\kappa^2)$. $\square$

5.   General networks

We now study the time-varying minimum spanning tree (TMST) problem on a general network. We will first examine its complexity in terms of strong NP-completeness, and then develop algorithms which can find, in pseudopolynomial time, approximate solutions.

5.1   Strong NP-hardness

It is well known that the classical minimum spanning tree problem is polynomially solvable (see, for example, Graham et al (1985)). We have shown in Sections 3 and 4 above that, the time-varying minimum spanning tree (TMST) problem on an arc series-parallel network or a reducible network is, however, NP-complete in the ordinary sense. In this section we will further show that the general TMST problem is NP-complete in the strong sense even if the underlying graph of N is a tree with $b(x,y,t) = b(x,y)$, or $c(x,y,t) = c(x,y)$ for any arc $(x,y) \in A$.

We will show that the *Minimum Set Cover (MSC) problem* is reducible to TMST.

Definition 2.13 MSC: *Given a set* $C = \{C_1, C_2, ..., C_m\}$ *of finite sets and a number* K_s, *does there exist a set cover* C' *such that* $|C'| \leq K_s$?

To study its complexity in terms of NP-completeness, we examine the decision version of the TMST problem as stated below.

Definition 2.14 TMST: *Given a time-varying network N and two integers k and κ, does there exist a spanning tree within the time limit κ such that its total cost is not greater than k ?*

We first examine the problem with the constraint that waiting at any vertex is not allowed. Our results are given in Theorem 2.7 and Theorem 2.8 below.

Theorem 2.7 *If waiting at any vertex is not allowed, then TMST is NP-complete in the strong sense, even if the underlying graph of N is a tree, and*

     (i) $c(x,y,t)$ are time-varying and $b(x,y,t) = b(x,y)$, $\forall (x,y) \in A$;

or

     (ii) $b(x,y,t)$ are time-varying and $c(x,y,t) = c(x,y)$, $\forall (x,y) \in A$.

Proof: Clearly, TMST is in NP. Now, we will prove that MSC reduces to TMST with no waiting allowed at any vertices. For each $u \in \bigcup_{i=1}^{m} C_i$, we create a vertex x to represent it in N. Moreover, add a source vertex s and a linking vertex v_0 in N. Use arcs (s, v_0) and (v_0, u) to connect these vertices and let

$$b(s, v_0, \kappa) = \infty, b(s, v_0, t) = 1, t = 0, 1, 2, ..., \kappa - 1$$
$$c(v_0, u, 0) = \infty, c(s, v_0, t) = 1, t = 0, 1, 2, ..., \kappa$$
$$b(v_0, u, t) = \begin{cases} 1 & \text{if } u \in C_t \\ \infty & \text{otherwise} \end{cases}$$

$$t = 1, 2, ..., \kappa$$

$$c(v_0, u, t) = 0, t = 1, 2, ..., \kappa$$

Let $k = K_s$ and $\kappa = m$. Finally, assume that at any $x \in V$, no waiting is allowed.

The network N defined above possesses the property that b are time-varying and c only depend on arc (x, y). Clearly, this reduction can be implemented in polynomial time (see Figure 2.12).

We now prove that a "yes" answer to MSC is equivalent to a "yes" answer to TMST.

If MSC has a set cover C' with $|C'| = l \leq K_s$, then a spanning tree with zero waiting times at any intermediate vertices can be constructed starting with the source vertex s. Without loss of generality, suppose $C_i \in C', 1 \leq i \leq l$ (note that $l \leq K_s \leq m$). Let $C_1' = C_1$ and $C_i' = C_i \backslash \bigcup_{j=1}^{i-1} C_j$, $2 \leq i \leq l$. Clearly, $C' = \bigcup^i C_i = \bigcup^i C_i'$. For each C_i', if $|C_i'| \neq 0$, we choose the path $P_i(s, v_0)$ starting at s at time $i - 1$ and arriving at v_0 at time i. If u is the element that occurs in C_i', we add

Figure 2.12. A network constructed from MSC

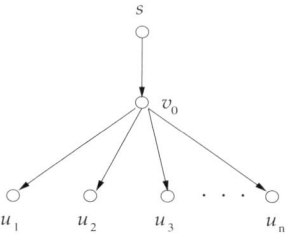

Figure 2.13. The spanning tree is splitted into subtrees

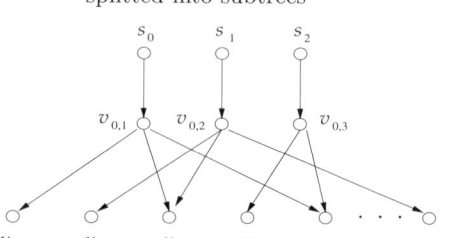

arc (v_0, u) to $P_i(s, v_0)$ with the starting time i and arriving time $i+1$ to form a subtree, denoted by T_i. According to our reduction, each vertex $u \in \bigcup_{i=1}^{m} C_i$ can be reached within time κ. It is obvious that the tree constructed by combining all T_i above is a spanning tree T within time $\kappa = m$ and the total cost is less than $k = K_s$ (see Figure 2.13).

If TMST has a spanning tree T of total cost not exceeding $k = K_s$, then the cost restriction guarantees that there exist less than k subtrees T_i which contain paths from s to each element $u \in \bigcup_{i=1}^{m} C_i$. Choose all elements u in T_i to form the set C_i'. Obviously, $C' = \bigcup_i C_i'$ contains all elements $u \in \bigcup_{i=1}^{m} C_i$, and C' is a set cover with $|C'| \le k = K_s$.

For the case $b(x, y, t) = b(x, y)$ and c are time-varying, we can modify the reduction as follows:

$$b(x, y, t) = 1, \forall (x, y) \in A, 0 \le t \le \kappa$$
$$c(s, v_0, \kappa) = \infty$$
$$c(v_0, u, t) = \begin{cases} 0 & \text{if } u \in C_t \\ \infty & \text{otherwise} \end{cases}$$
$$t = 0, 1, ..., \kappa$$

All other analyses remain unchanged. This completes the proof. $\square$

A problem is said to be in the class of APX, if it has a constant-error approximation algorithm, i.e., an algorithm that can find, in polynomial time, an approximate solution with an error bound β, where $\beta > 1$ is a fixed constant (Note that the error bound is defined as follows: Let ζ^* and ζ^0 be the optimal and approximate solutions, respectively. Then, ζ^0 is said to have an error bound β if $\zeta^0/\zeta^* \le \beta$). Lund and Yannakakis (1993) indicate that MSC is not in the class of APX; i.e., to find a constant-factor approximation algorithm for MSC is at least as hard as to prove P=NP. The reduction we constructed above implies that the TMST problem is also not in the class of APX. This gives us the following result.

Theorem 2.8 *Consider the TMST problem with no waiting allowed at any vertex. There is no constant-error polynomial-time approximation algorithm for the TMST problem unless P=NP, even if the underlying graph of N is a tree, waiting time at any vertex must be zero, and one of the two parameters, b or c, is time independent.*

We now consider the situation where waiting at any vertex is arbitrarily allowed (Namely, waiting at any vertex is subject to no constraints). We will establish its NP-completeness using another reduction from MSC. Note that in the case where waiting at any vertex is prohibited, we can show its NP-completeness even when the underlying graph of N is a tree. This is however not extendible to the case with waiting at any vertex being arbitrarily allowed. The reduction analysis is now built on a network that is not a rooted tree; see below.

Theorem 2.9 *The TMST problem where waiting at any vertex is arbitrarily allowed is NP-complete in the strong sense, even if $c(x,y,t) = c(x,y)$ and $b(x,y,t) = b(x,y)$, $\forall (x,y) \in A$.*

Proof: For any given instance of MSC, we construct an instance of TMST as follows: For each $C_i \in C$, create a pair of vertices x_i and x_i' ($1 \leq i \leq m$). For each element $u_j \in \bigcup C_i$, create a vertex x_{m+j}. These vertices together with an extra vertex, the source vertex s, compose the vertex set V. Create arcs (s, x_i), (s, x_i'), and (x_i', x_i), $1 \leq i \leq m$, as shown in Figure 2.14. For each vertex x_{m+j}, create an arc (x_i, x_{m+j}) if $u_j \in C_i$ (see Figure 2.14). These arcs compose the arc set A. Let

$$b(s, x_i, t) = b(s, x_i', t) = b(x_i', x_i, t) = b(x_i, x_{m+j}, t) = 1,$$
$$1 \leq i \leq m, 1 \leq j \leq n, 0 \leq t \leq \kappa$$
$$c(s, x_i, t) = 1, c(s, x_i', t) = c(x_i', x_i, t) = 0, 1 \leq i \leq m, 0 \leq t \leq \kappa$$
$$c(x_i, x_{m+j}, t) = 0, 1 \leq i \leq m, 1 \leq j \leq n, 0 \leq t \leq \kappa$$

Finally, let $k = K_s$, $\kappa = 2$. The network $N(V, A, b, c)$ as defined above has the property that both c and b are time independent (Note that we do not have any restriction on the waiting time at any vertex). In what follows, we prove that a "yes" answer to MSC is equivalent to a "yes" answer to TMST.

If MSC has a set cover C' with $|C'| = l \leq K_s$, then a spanning tree with waiting times at its vertices can be constructed as follows: Without loss of generality, assume $C_i \in C'$, $1 \leq i \leq l$ (note that $l \leq K_s \leq m$), and let $C_1' = C_1$ and $C_i' = C_i \backslash \bigcup_{j=1}^{i-1} C_j$, $2 \leq i \leq l$. For each C_i', if $|C_i'| \neq 0$, we choose the path $P_i(s, x_i)$ with $\tau(s) = 0$ and $\alpha(x_i) = 1$. If u_j is the element in C_i', we add arc (x_i, x_{m+j}) to $P_i(s, x_i)$ with $\tau(x_i) = 1$

Figure 2.14.   A time-varying network created from MSC

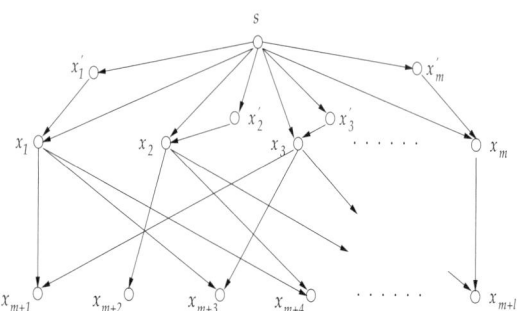

and $\alpha(x_i) = 2$. When there are more than one arc added to $P_i(s, x_i)$, we have a subtree, denoted by T_i. Since $\cup_{i=1}^{l} C_i = \cup_{i=1}^{l} C_i' = \cup_{i=1}^{m} C_i$, each vertex x_{m+j} can be reached within time 2. For those $C_i \notin C'$, we choose the path $P(s, x_i) = (s, x_i', x_i)$ with $\tau(s) = 0$, $\alpha(x_i') = \tau(x_i') = 1$ and $\alpha(x_i) = 2$. It is obvious that all vertices in V can be reached within time 2. Combining all T_i and P_i' we obtain a spanning tree T within time $\kappa = 2$. Since the cost of T_i is equal to 1 and the cost of P_i' is 0, we have $\zeta(T) \leq l \leq k = K_s$.

Now, if TMST has a spanning tree $T(2)$ with $\zeta(T(2)) \leq k \leq K_s$, then the cost and transit time in the problem constructed above guarantee that there exist $l \leq k$ subtrees, which contain path(s) from s to each vertices x_{m+j} with $\alpha(x_{m+j}) = 2$. Choose those sets C_i if x_{m+j} appears in T_i. Let $C' = \bigcup_{i=1}^{l} C_i$. Since $T(2)$ contains all vertices x_{m+j}, we have all elements $u_j \in \bigcup_{i=1}^{l} C_i$. Thus C' is a set cover with $|C'| \leq k = K_s$. In summary, we complete the proof.   □

Again, from the definition of the class of APX, we have:

Theorem 2.10 *Consider the TMST problem where waiting at any vertex is arbitrarily allowed. There is no constant-error polynomial-time approximation algorithm for the problem unless P=NP, even if for any arc $(x, y) \in A$, $b(x, y, t) = b(x, y)$ and $c(x, y, t) = c(x, y)$.*

Finally, we consider the general TMST problem where waiting time at any vertex may be constrained by an arbitrary function. Since this contains the situation where waiting time at a vertex must be zero as its special case, Theorem 2.11 and Theorem 2.12 below follow immediately from Theorem 2.7 and Theorem 2.8, respectively.

Theorem 2.11 *The general TMST problem is NP-complete in the strong sense, even if the underlying graph of N is a tree and*
   (i) $c(x, y, t)$ are time-varying and $b(x, y, t) = b(x, y)$, $\forall (x, y) \in A$; or

(ii) $b(x, y, t)$ are time-varying and $c(x, y, t) = c(x, y)$, $\forall (x, y) \in A$.

Theorem 2.12 *There is no constant-error polynomial-time approximation algorithm for the general TMST problem unless P=NP, even if the underlying graph of N is a tree and one of the two parameters, b or c, is time independent.*

5.2   Heuristic algorithms

We now describe our algorithms that can find approximate solutions in pseudopolynomial time. Basically, our algorithms consist of two main steps: Firstly, for each vertex $x \in V$, we identify the shortest path from s to x of time at most κ. Putting all these paths together we obtain a path-induced subnetwork of N. Then, we remove redundant arcs from this subnetwork so as to obtain a spanning tree of N. We will show that the resulted tree, denoted as $T_A(\kappa)$, is an approximate solution for the general TMST problem, while the total time required to construct $T_A(\kappa)$ is pseudopolynomial.

In order to apply the algorithms in Chapter 1 to find the shortest paths, in the following we will limit the problem to be solved to the one where waiting time at vertex x, for $x \in V$, is bounded above by u_x. Note that when $u_x = 0$ for all $x \in V$, we have the case with no waiting at any vertex being allowed, while if $u_x = \infty$ for all $x \in V$, we have the case where waiting at x is arbitrarily allowed.

5.2.1 Finding a shortest path

The definition below defines the return function of the dynamic program to be introduced below.

Definition 2.17 *Let $d_b(y, t)$ be the cost of a shortest path from s to y of time exactly t. If such a path does not exist, let $d_b(y, t) = \infty$.*

We rewrite Algorithm TSP-BW as a procedure below:

> **Procedure DP**
> **Begin**
> **Initialize:** $d_b(s, 0) := 0$, and $d_b(x, 0) := \infty$, $\forall x \neq s$; $d_b(x, t) := \infty$, $\forall x$ and $\forall t > 0$; $Heap_x := \{d_b(x, 0)\}$ and $d_b^m(x, 0) := d_b(x, 0)$, $\forall x$;
> > **Sort** all values $u + b(x, y, u)$ for all $u = 0, 1, ..., \kappa$ and for all arcs $(x, y) \in A$;
> > **For** $t = 1, ..., \kappa$ **do**
> > > **For** every arc $(x, y) \in A$ **do** $\mathcal{R}_b(x, y, t) := \infty$;
> > > **For** all arcs $(x, y) \in A$ and all u_D such that $u_D + b(x, y, u_D) = t$
> > **do**

$$\mathcal{R}_b(x,y,t) := \min\{\mathcal{R}_b(x,y,t), d_b^m(x,u_D) + c(x,y,u_D)\};$$

For every vertex y **do** $d_b(y,t) := \min_{\{x|(x,y)\in A\}} \mathcal{R}_b(x,y,t)$;
For every vertex y **update** the heap as follows
  Insert-heap$_{(y)}$ $d_b(y,t)$;
  If $t > u_y$ **then delete-heap**$_{(y)}$ $d_b(y, t - u_y - 1)$;
For every vertex y **do**
  $u_A := $ **Minimum-heap**$_{(y)}$;
  $d_b^m(y,t) := d_b(y,u_A)$;
For every y **do** $d_b^*(y) := \min_{0\le t\le \kappa} d_b(y,t)$;
End.

5.2.2 Removing redundant arcs

After applying the algorithm above to find the shortest paths between the root s and each $x \in V$, we can obtain a path-induced subnetwork of N by combining all these paths together. The next main step of our approach is to remove those redundant arcs from the path-induced subnetwork. We will perform the following vertex/arc deleting operations:

(i) *Deleting shared intermediate vertices.* If a vertex x appears in two paths P_i and P_j, and if $[\alpha_j(x), \tau_j(x)] \subseteq [\alpha_i(x), \alpha_i(x) + u_x]$, then arc (y, x), where y is the predecessor of x in P_j, is redundant, since all successors of x in P_j can be reached from s through the section $P_i(s, x)$ in path P_i and the waiting time constraint is not violated. Therefore we can delete arc (y, x). Then, the original path P_j ends at the vertex y, and a new path is created which consists of two sections: the section $P_i(s, x)$ in path P_i, and the section starting from x in the original path P_j with the departure time $\tau_j(x)$. Repeat this operation until there are no shared intermediate vertices. Then go to Operation (ii) next.

(ii) *Deleting redundant end vertices.* Let x_j^o denote the end vertex of path P_j. If x_j^o appears in another path, or in P_j as an intermediate vertex, then delete the end vertex x_j^o in P_j as well as its adjacent arc, and the predecessor of x_j^o in P_j becomes the new end vertex. Repeat this operation until the whole path is eliminated or the end vertex of P_j does not appear in any other paths or in P_j as an intermediate vertex.

By Definition 2.2, the path-induced subnetwork N' generated after performing Operations (i) and (ii) above is a spanning tree. In what follows, we will use two procedures, DSIV (Deleting Shared Intermediate Vertices) and DREV (Deleting Redundant End Vertices) to realize these two operations respectively.

(1) Procedure DSIV

Procedure DSIV is used to delete the shared intermediate vertices among paths. To reduce the time requirement, the procedure uses a 3-dimensional array $e(x, t, i)$. If vertex x appears in path P_j with arrival time $\alpha_j(x)$ and departure time $\tau_j(x)$, then $e(x, t, 1) = j$ and $e(x, t, 2) = \tau_j(x)$, where $t = \alpha_j(x)$. Initially, they are set to zero. The procedure contains the following two basic steps:

(i) If there are more than one path which include x with the same arrival time $\alpha(x)$, then keep the path P_{j^0} that has the latest departure time $\tau_{j^0}(x)$. Delete all arcs (y_i, x) in path P_i, where y_i is the predecessor of x in path P_i and $i \neq j^0$.

(ii) Check array e. For each vertex x, if there exist u and t such that $t < u$ and $e(x, u, 2) \leq t + u_x$ (this means that there are two paths P_i and P_j with $e(x, t, 1) = i$, $e(x, u, 1) = j$, $t = \alpha_i(x)$, $u = \alpha_j(x)$ and $e(x, u, 2) = \tau_j(x)$, which satisfies $[\alpha_j(x), \tau_j(x)] \subseteq [\alpha_i(x), \alpha_i(x) + u_x]$), then arc (y_j, x) can be deleted in P_j.

After completing these two steps, all shared intermediate vertices will be deleted.

> **Procedure DSIV**
> **Begin**
> **For** $x \in V \backslash \{s\}$ and $t - 0, 1, ..., \kappa$ **do** $e(x, t, 1) := e(x, t, 2) := 0$;
> **For** each path P_j $(j = 1, ..., n - 1)$ and each $x \in V(P_j)$ **do**
>   **If** $e(x, \alpha(x), 1) = 0$ **then** $e(x, \alpha(x), 1) := j$, $e(x, \alpha(x), 2) :=$ $\tau(x)$;
>   **Else If** $\tau(x) \leq e(x, \alpha(x), 2)$ **then** delete arc (y_j, x) in P_j;
>   **Else** let $i := e(x, \alpha(x), 1)$, delete arc (y_i, x) in P_i, $e(x, \alpha(x), 1) := j$, $e(x, \alpha(x), 2) := \tau(x)$;
>   **For** each $x \in V \backslash \{s\}$ **do**
>   **Let** $\alpha := 0$;
>   **For** $t = 0, 1, ..., \kappa$ **do**
>   **If** $e(x, t, 1) = 0$ **then** $\alpha := t$;
>   **Else If** $e(x, t, 2) \leq \alpha + u_x$ **then** let $i := e(x, \alpha(x), 1)$, delete arc (y_i, x) in P_i, $\alpha_i(x) := \alpha$, $e(x, t, 1) := e(x, t, 2) := 0$;
> **End**.

The time complexity of the procedure can be analyzed as follows. The first and the third For-do loops need $O(n\kappa)$ time. The second loop also takes $O(n\kappa)$ time since each path P_j contains at most κ vertices. Therefore, the total running time of the procedure is bounded by $O(n\kappa)$.

(2) Procedure DREV

This procedure is used to delete the redundant end vertices of the paths. The basic idea is to set up a counter $num(x)$ for each vertex x to record the number of its occurrences in all paths. Then, when we check whether an end vertex x of a path appears in another place, we only need to check $num(x)$. If $num(x) > 1$, it is clear that x must appear in another place and so we can delete it from the path and then decrease $num(x)$ by 1.

> **Procedure DREV**
> **Begin**
> $num(0) := 0$;
> **For** each $x \in V \backslash \{s\}$ **do** $num(x) := 0$;
> **For** each $x \in V \backslash \{s\}$ and $t = 0, 1, ..., \kappa$ **do**
> **If** $e(x, t, 1) > 0$ **then** $num(x) := num(x) + 1$;
> **For** each path P_j **do**
> **Identify** the end vertex x_j^o of P_j;
> **While** $num(x_j^o) > 1$ **do**
> **Let** $num(x_j^o) := num(x_j^o) - 1$;
> **Delete** x_j^o in P_j;
> **End** *while*;
> **End**.

The procedure DREV checks each vertex in all paths. Since there are at most $n-1$ paths (because there are at most $n-1$ end vertices) in $T_A(\kappa)$ and each path contains at most κ vertices (because the transit time b is a positive integer and the arrival time at a vertex must be greater than that at its predecessor), the procedure DREV needs at most $O(n\kappa)$ time.

5.2.3 The algorithm A-TMST

Our algorithm can be described as follows.

> **Algorithm A-TMST**
> **Begin**
> **Call** Procedure DP to obtain the shortest path $P_j(s, x)$, from s to each vertex $x \in V \backslash \{s\}$, for $1 \leq j \leq n - 1$;
> **If** there exists a path $P(s, x)$ with $d^*(x) = \infty$ **then** let $\zeta(T_A(\kappa)) := \infty$ and stop;
> **Call** procedure DSIV to delete the shared intermediate vertices;
> **Call** procedure DREV to delete the redundant end vertices;
> **Combine** all paths that remain to generate an path-induced subnetwork $T_A(\kappa)$;
> **End**.

An approximate solution that has an error bound f is called *an f-approximate solution*. Let L denote all leaves in $T_A(\kappa)$ obtained by A-TMST and $l = |L|$. Then, we have

Theorem 2.13 *Algorithm A-TMST can find, in at most $O(\kappa(m + n\log\kappa))$ time, an l-approximate solution for TMST. Moreover, l is the best possible bound for the algorithm A-TMST.*

Proof: Clearly, $T_A(\kappa)$ obtained by the algorithm is a spanning tree of N within time κ, according to Definition 2.2.

Let us first consider the time requirement of A-TMST. The running time required by Procedure DP is $O(\kappa(m + n\log\kappa))$. Since the procedures DSIV and DREV each need only $O(n\kappa)$ time, the total running time of A-TMST is thus bounded above by $O(\kappa(m + n\log\kappa))$.

We now analyze the error bound of the approximate solution $T_A(\kappa)$. Recall that l is the number of leaves in $T_A(\kappa)$. It is not hard to see that

$$\zeta(T_A(\kappa)) \le \sum_{x\in L}\zeta(P^*(s,x)) \le \sum_{x\in L}\zeta(P(s,x)) \le l\zeta(T(\kappa)),$$

where $T(\kappa)$ is the optimal spanning tree and $P(s,x)$ is the path in $T(\kappa)$ from s to x, while $P^*(s,x)$ is the shortest path from s to x obtained by Procedure DP.

Figure 2.15. A network for error bound analysis

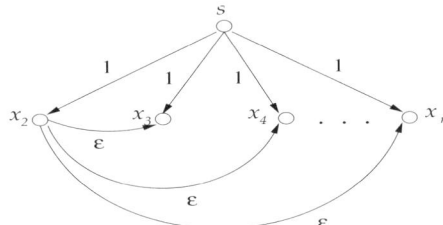

To show that the bound l is the best achievable by Algorithm A-TMST, we consider the following special class of the time-varying networks N, and illustrate that the approximate solution obtained by applying Algorithm A-TMST on such networks will achieve the bound l. Let $N_\varepsilon = (V_\varepsilon, A_\varepsilon, b_\varepsilon, c_\varepsilon)$ (see Figure 2.15), where $V_\varepsilon = \{x_1(= s), x_2, ..., x_n\}$, $A_\varepsilon = \{(s, x_i), 2 \le i \le n, (x_2, x_j), 3 \le j \le n\}$, and

$$b_\varepsilon(s, x_i, t) = 1, c_\varepsilon(s, x_i, t) = 1, 2 \le i \le n, 0 \le t \le \kappa$$
$$b_\varepsilon(x_2, x_j, t) = \varepsilon, c_\varepsilon(x_2, x_j, t) = 1, 3 \le j \le n, 0 \le t \le \kappa.$$

Note that when $\kappa \ge 2$, the optimal spanning tree $T(\kappa)$ is Figure 2.16 with $\zeta(T(\kappa)) = 1 + (n-2)\varepsilon$. The solution of A-TMST, $T_A(\kappa)$, is Figure

Figure 2.16. The optimal solution   *Figure 2.17.* The solution obtained by A-TMST

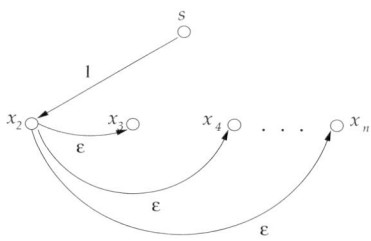   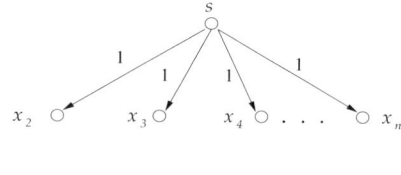

2.17 with $\zeta(T_A(\kappa)) = n - 1$. Letting ε tend to zero, we have

$$\frac{\zeta(T_A(\kappa))}{\zeta(T(\kappa))} = \frac{n - 1}{1 + (n - 2)\varepsilon} \xrightarrow{\varepsilon \to 0} n - 1 = l$$

where l is the number of leaves in $T_A(\kappa)$. Therefore, we complete the proof. □

Remark. Note that the time complexity of the algorithm A-TMST is dominated by that of the procedure to find shortest paths.

As for the error bound of the algorithm A-TMST, can we further improve it? Theorem 2.13 indicates that the bound l is the best possible, in general. Nevertheless, in some special situations where certain structure/conditions are satisfied, we may get a better error bound. In the following we discuss such a case.

5.3   The error bound of the heuristic algorithms in a special case

We will show here that the error bound of the algorithm A-TMST can be improved for a type of multi-period networks. Such a network possesses a multi-period structure, with each period starting from a common source vertex; see Figure 2.18 (Note that the parameters, such as the transit times and the transit costs on the arcs, can be time varying).

Let N_i denote the sub-network of a period in N, $1 \le i \le k$; see Figure 2.18. Furthermore, let $L(N_i)$ denote the leaves of $T_A(N_i)$, where $T_A(N_i)$ is the sub-tree of $T_A(\kappa)$ covering N_i, and let $l_i = |L(N_i)|$ $(1 \le i \le k)$. Then, we have

Corollary 2.1 *The approximate solution obtained by the algorithm A-TMST has an error bound $f = max\{l_1, l_2, ..., l_k\}$, if N is a job scheduling network and*

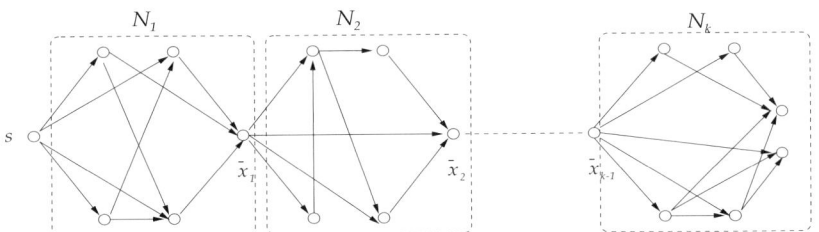

Figure 2.18. A network with multi-period structure

(i) waiting at any vertex is allowed without any restriction, and
(ii) both $b(x,y,t)$ and $c(x,y,t)$ are nonincreasing over time t.

Proof: We use the same notation as in the proof of Theorem 2.13. First, we will show the following inequalities:

$$\sum_{x\in L(N_i)} \zeta(P^*(\bar{x}_{i-1},x)) \le \sum_{x\in L(N_i)} \zeta(P(\bar{x}_{i-1},x)), \quad \text{for } 2\le i\le k. \quad (2.1)$$

Note that, due to the structure of the network, each path $P^*(s,x)$ and $P(s,x)$, $x\in V(N_i)$, must pass through the vertex $\bar{x}_{i-1}$ ($2\le i\le k$). Moreover, since waiting at a vertex is allowed, we claim that each path $P^*(s,x)$ and $P(s,x)$ ($x\in L(N_i)$) must contain the same subpath $P^*(s,\bar{x}_{i-1})$ and $P(s,\bar{x}_{i-1})$ ($2\le i\le k$). Otherwise, suppose that there are two paths, say $P(s,x_1)$ and $P(s,x_2)$, which have different subpaths $P_1(s,\bar{x}_{i-1})$ and $P_2(s,\bar{x}_{i-1})$, where x_1 and $x_2\in L(N_i)$. Then, let $\alpha_1(\bar{x}_{i-1})$ and $\alpha_2(\bar{x}_{i-1})$ be the arrival times at vertex $\bar{x}_{i-1}$ on $P(s,x_1)$ and $P(s,x_2)$, respectively. Without loss of generality, suppose $\alpha_1(\bar{x}_{i-1}) < \alpha_2(\bar{x}_{i-1})$. Then we can delete the arc $(y,\bar{x}_{i-1})$ in $P_2(s,\bar{x}_{i-1})$, where y is the predecessor of $\bar{x}_{i-1}$ in P_2. Denote the new tree as $T^0(\kappa)$. Clearly, we have $\zeta(T^0(\kappa)) < \zeta(T(\kappa))$. This contradicts the fact that $\zeta(T(\kappa))$ is optimal.

Now, let $\alpha^*(\bar{x}_{i-1})$ and $\alpha(\bar{x}_{i-1})$ be the arrival times at vertex $\bar{x}_{i-1}$ in $P^*(s,x)$ and $P(s,x)$, respectively. Consider two cases:

(i) $\alpha^*(\bar{x}_{i-1}) \le \alpha(\bar{x}_{i-1})$. The inequalities (1) hold clearly since any shortest path $P^*(\bar{x}_{i-1},x)$ starting from time $\alpha^*(\bar{x}_{i-1})$ is shorter than that starting from time $\alpha(\bar{x}_{i-1})$, and also shorter than the path $P(\bar{x}_{i-1},x)$.

(ii) $\alpha^*(\bar{x}_{i-1}) > \alpha(\bar{x}_{i-1})$. Because both $b(x,y,t)$ and $c(x,y,t)$ are nonincreasing functions of time t, any shortest path $P^*(\bar{x}_{i-1}),x)$ starting from time $\alpha^*(\bar{x}_{i-1})$ is shorter than that starting from time $\alpha(\bar{x}_{i-1})$, and also shorter than path $P(\bar{x}_{i-1}),x)$.

In summary, we prove the inequalities (1). Now, it follows from (1) that

$$\zeta(T_A(\kappa))$$

$$\leq \sum_{x \in L(N_1)} \zeta(P^*(s,x)) + \sum_{x \in L(N_2)} \zeta(P^*(\bar{x}_1,x)) + ... + \sum_{x \in L(N_k)} \zeta(P^*(\bar{x}_{k-1},x))$$

$$\leq \sum_{x \in L(N_1)} \zeta(P(s,x)) + \sum_{x \in L(N_2)} \zeta(P(\bar{x}_1,x)) + ... + \sum_{x \in L(N_k)} \zeta(P(\bar{x}_{k-1},x))$$

$$\leq l_1\zeta(T_{N_1}(\kappa)) + l_2\zeta(T_{N_2}(\kappa)) + ... + l_k\zeta(T_{N_k}(\kappa))$$

$$\leq \max\{l_1, ..., l_k\}\zeta(T(\kappa))$$

where $T_{N_i}(\kappa)$ is the sub-spanning tree of $T(\kappa)$ on sub-network N_i, $1 \leq i \leq k$. That is,

$$\frac{\zeta(T_A(\kappa))}{\zeta(T(\kappa))} \leq \max\{l_1, ..., l_k\}$$

Therefore, we complete the proof.   □

5.4   An approximation scheme for the problem with arbitrary waiting constraints

For the time-varying minimum spanning tree problem on the general network but under arbitrary waiting time constraints, we have another approximate scheme, which can solve the problem more efficiency. The scheme consists of two main steps: Firstly, for a given general time-varying network, create a spanning graph of N, denoted by N', which contains no subgraph homomorphic to K_4. Then, we apply Algorithm TMST-RN on N'. Let T' be the optimal solution obtained by the algorithm. Clearly, T' is an approximate solution of the original network N. Both of these two steps can be implemented in pseudopolynomial time.

5.4.1   Creating a spanning reducible network

Remind that $A(N)$ is the arc set of the original network N and s is the root of N. The basic idea of creating the spanning reducible network is:

(i) Let Q be a vertex set. Initially, set $Q = \{s\}$. Denote $dt(x)$ as the earliest possible departure time at x. Let $dt(s) = 0$.

(ii) Pick up a vertex, say, x, from Q. Let $adj(x) = \{(x,y)|(x,y) \in A(N), (x,y)$ is unchecked$\}$, which denotes the all unchecked adjacent arcs of x, and sort it in nonincreasing order in terms of the value of $c(x,y,dt(x))$. Do the following repeatedly till $adj(x)$ becomes empty:

(a) Pick up the first arc (x,y) in $adj(x)$ and add it in N'. Check whether N' contains a subgraph homomorphic to K_4 or not. If the answer is "yes", remove (x,y) from N'; Otherwise, leave (x,y) in N' (still called the new network obtained as N'). If $y \notin Q$ then let $Q = Q + y$.

(b) Delete (x,y) from $adj(x)$ and (x,y) is said to be checked.

(iii) Do (ii) repeatedly till Q becomes empty.

Clearly, N' is a spanning reducible network of N with edges as many as possible.

Now we give the algorithm as below.

> **Algorithm TMST-A**
> **Begin**
> Set $E(N') = \emptyset$ and $V(N') = \emptyset$. Let $Q = \{s\}$, $dt(s) = 0$;
> **While** $Q \neq \emptyset$ **do**;
>   Select a vertex x from Q;
>   Sort $adj(x)$ in nonincreasing order on the value of $c(x, y, dt(x))$;
>   **While** $adj(x) \neq \emptyset$ **do**;
>   Pick up the first edge $(x, y) \in adj(x)$;
>   **If** N' contains a subgraph homomorphic to K_4, then discard (x, y);
>   **Else** let $A(N') = A(N') + (x, y)$ and $V(N') = V(N') + y$ if $y \notin V(N')$;
>   Let $adj(x) = adj(x) \backslash (x, y)$;
>   **End** *while*;
>   Let $Q = Q \backslash x$;
> **End** *while*;
> **End**

Clearly, N' obtained by the algorithm is a reducible network of N.

<u>Theorem 2.14</u> *Algorithm TMST-A can be implemented in $O(m \cdot \max\{m, n\})$.*

Proof: The initialization can be done in constant time. Since Q will contain at most n vertices, the first "while" loop will be performed in at most n times. Sorting $adj(x)$ needs $O(d(x) \log d(x)) = O(d(x) \log m)$, where $d(x)$ is the degree of vertex x. Since we need do this step for all vertices in N, the total number of performances is $\sum_{x \in V(N)} d(x) \log m \leq m \log m$. During the second "while" loop, checking whether N' contains a subgraph homomorphic to K_4 needs $O(\max\{m, n\})$ (see Liu and Geldmacher 1980). As we need do this for all edges in $adj(x)$ and for all vertices x, it needs $O(m \cdot \max\{m, n\})$ time. In summary, the total running time of Algorithm TMST-A is bounded above by $O(m \cdot \max\{m, n\})$. $\square$

5.4.2 Numerical experiments

We test the approximate algorithm on PC. Table 2.6 illustrate our numerical results. The size of problem is listed on the first column with the number of vertices n, the number of edges m, and the time duration T we considered, respectively. All the transit time b, cost c

are generated randomly. In the right column of Table 2.6, m' is the number of spanning reducible network created on the original network by Algorithm TMST-A. For example, the problem 1 has 30 vertices, 80 edges with time duration $T = 30$. The edges of the spanning reducible network N' is 38. The cost of the minimum spanning tree of N' is 31, and use 2 seconds CPU time. As a comparison, we apply Monte Carlo Method to the same problem. Generate 1000 minimum spanning trees for the original network N and choose the best one, which has the cost 70 and costs 170 seconds CPU time. All numerical experimental results show that our approximate algorithm is much batter than Monte Carlo Method.

Table 2.6. The numerical experimental results

Problem size			Monte Carlo method			Approximate algorithm		
n	m	T	repeating times	cost	CPU (sec.)	m'	cost	CPU (sec.)
30	80	30	1000	70	170	38	31	2
50	120	50	5000	113	454	62	49	5
80	260	60	5000	202	515	98	80	36
100	300	80	5000	265	631	127	99	69
150	400	100	5000	394	948	176	149	204
200	500	100	5000	527	1282	228	202	366

6.   Additional references and comments

Applications of MST models have been extensively studied in the literature. These include physical systems design (Prim (1957); Loberman et al (1957); Dijkstra (1959)), network design (Magnanti et al (1984)), optimal message passing (Abdel-Wahab et al (1997); Prim (1957)), pattern classification (Dude et al (1973)), image processing (Osteen et al (1974); Xu et al (1997)), and network reliability analysis (Van Slyke et al (1972)). More references on applications of the MST problem can be found in Graham et al (1985). Efficient algorithms for the MST problem include those proposed by Kruskal (1956); Prim (1957), and Dijkstra (1959). The directed MST problem can be solved by an algorithm proposed by Edmonds (1965).

Solomom (1986) considers the situation where there is a time window associated with each vertex. A transit time is needed to traverse an arc, and any vertex must be visited within its time window. The problem is to find a minimum spanning tree to cover all the vertices under the time-window constraints. Solomom proves that this problem is NP-hard, and presents a greedy algorithm and an insertion algorithm, which may generate approximate solutions for the problem. By appropriating setting the parameters in our time-varying MST model, we can show that Solomom's problem is a special case under our framework.

Chapter 3

TIME-VARYING UNIVERSAL MAXIMUM FLOW PROBLEMS

1. Introduction

The maximum flow (MF) problem aims to find a solution that can send the maximum flow from one vertex (the source) to another vertex (the sink), under the constraint that the capacities of the arcs and vertices in the network are all satisfied. Most MF models considered in the literature are static, in which the capacities of both the arcs and vertices are assumed to be constant. In practical situations, it is easy to see many time-varying MF problems. For instance, consider a transportation network in which several cargo-transportation services are available among a number of cities. Each of them may take a certain time to travel from one city to another, with a limited cargo-transporting capacity. Moreover, the travel time and the capacity for each transportation service are season dependent. A question often asked is: what is the maximum flow that can be sent between two specific cities within a certain duration T? This is a time-varying maximum flow problem. Its solution is important, particularly to the planning of the network.

An extension of the classical static problem is the maximal dynamic flow model formulated and solved by Ford et al (1962), where the transit time to traverse an arc is taken into consideration. Nevertheless, their model still assumes that attributes in the problem, including arc capacities and transit times, are time independent. Ford and Fulkerson have developed an efficient procedure to find the optimal solution for their model , which first finds the static flow from the source to the sink, and then develops a set of temporally repeated flows, with the optimal flow decomposed into a set of chain flows. A further extension is studied by Halpern (1979), where arc capacities vary over time and storages at intermediate vertices may be prohibited at some times. When the time is considered as a variable taking discrete values, both these problems can be solved by constructing an equivalent, static time-expanded network (see, e.g., Minieka (1978)).

Gale (1959) introduces the concept of *universal maximal dynamic flow* within a given duration T, which is defined as such a flow solution that remains optimal when the deadline T is truncated to any smaller t, $0 \leq t \leq T$. More specifically, for a time-varying network, the concept of universal maximal dynamic flow is defined as follows. Let $\lambda(T)$ be the optimal schedule to send flows under the time limit T, and $\lambda(t)$ the sub-schedule of $\lambda(T)$ which sends those flows that arrive at the sink no later than t, where $t \leq T$. Then, for any $t \leq T$, if $\lambda(t)$ remains the optimal schedule under the time limit t, then $\lambda(T)$ is *a universal maximum flow under the time limit T*. Note that the solution derived by Ford and Fulkerson's algorithm does not necessarily give the universal maximum flow. Minieka (1973) and Wilkinson (1971) have independently modified Ford and Fulkerson's algorithm to produce a universal maximal flow for the network studied by Ford and Fulkerson.

It is important to know the universal maximum flow that can be sent through a time-varying network. For example, in the cargo-transportation network described above, suppose that the source node is the base of a manufacturing firm while the sink is the primary market for the manufacturer's products. Further, suppose that the manufacturer produces different products in different seasons, which are perishable and therefore must be sent to the destination within different time limits. In such an environment, the information on the universal maximum flow will be useful for the manufacturer to decide on the maximum amount of his production at different times. The information will also be helpful for him to plan the routes and means to transport his products.

In this chapter, we will study the universal maximal flow model in a time-varying network, where transit times, arc capacities and vertex capacities are all time-varying. We will introduce, in Section 2, some basic definitions. The complexity of the problem with respect to NP-completeness will also be discussed in this Section. Section 3 studies the concept of residual network, an important tool to be used to decompose the problem into a set of subproblems. The well-known *Max-Flow Min-Cut Theorem* will be generalized, in Section 4, to the time-varying model. A condition to ensure an f-augmenting path to be feasible will be presented in Section 5, while algorithms that can generate the optimal universal maximum flow will be developed in Section 6. Finally, some concluding remarks will be given in Section 7.

2. Definition and problem formulation

Let $N(V, A, b, l)$ be a time-varying network, where V, A and b are as defined before. The parameter $l(x, y, t)$ is defined as the capacity of the arc (x, y) at time t, which represents the maximum amount of flow that can travel over arc (x, y) when the flow departs from x at time t, and the parameter $l(x, t)$ is defined as the capacity of the vertex x, which represents the maximum amount of flow that can stay (wait) at x during the time period $[t, t+1)$. Both $b(x, y, t)$ and $l(x, y, t)$ are functions of the departure time t at x, where $t = 0, 1, ..., T$, and $T > 0$ is a given number. The vertex capacity $l(x, t)$ is a function of the time t when the flow arrives at the vertex x.

The vertex capacity $l(x, t)$ applies only to the flow that waits at the vertex. In other words, it does not apply to the flow that passes through the vertex without waiting. This models, for example, the situation where some commodity has to wait at a port x for a while, and $l(x, t)$ represents the inventory capacity of x at time t. We assume that l is a nonnegative integer and the transit time b is a positive integer. We further assume that two vertices, s and ρ, are the source vertex and the sink vertex, respectively.

Without ambiguity, we let $f(x, y, \tau)$ be the value of the flow departing at time τ to traverse the arc (x, y), and $f(\lambda, t)$ the total flow value under the solution λ, which specifies when and how to send flows from the source s to the sink ρ within the time limit $t \leq T$. Then, it is clear that

$$f(\lambda, t) = \sum_{(x, \rho) \in A, \tau + b(x, \rho, \tau) \leq t} f(x, \rho, \tau).$$

Moreover, it is clear that $f(\lambda, T)$ is the value of flows sent from s to ρ no later than the time limit T.

The time-varying maximum flow problem is to find a solution λ such that $f(\lambda, T)$ is maximized. The *time-varying universal maximum flow* (TVUMF) problem is to find a solution λ^* such that $f(\lambda^*, t)$ remains the optimal solution for the time-varying maximum flow problem for any $0 \leq t \leq T$.

Note that there is an interesting distinction between the problem of maximum flow and the problem of universal maximum flow. This can be illustrated by the following example.

Example 3.1

Consider a time-varying network N as shown in Figure 3.1, where $T = 5$, and

$$b(s, g, t) = 1, b(g, h, t) = 1, b(h, \rho, t) = 1, b(s, h, t) = 3, b(g, \rho, t) = 3,$$

$l(x, y, t) = 1, l(x, t) = 0$, for all x, y, and $0 \le t \le T$.

Figure 3.1. A time-varying maximum flow problem

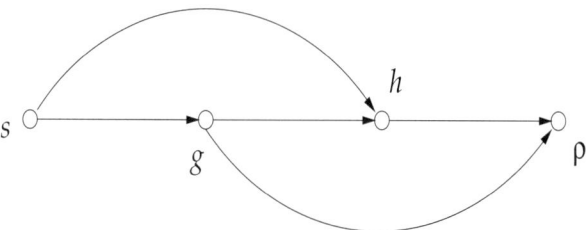

Clearly, one can have such a solution λ: At time $t = 0$, one unit of flow is sent along the path $P_1 = (s, h, \rho)$ and one unit of flow is transmitted along the path $P_2 = (s, g, \rho)$ (which all arrive at vertex ρ at time 4); At time $t = 1$, one unit of flow is sent along each of the two paths (which arrive at vertex ρ at time 5). Consequently, $f(\lambda, 4) = 2$ and $f(\lambda, 5) = 4$. One can verify (e.g., using the Time-Varying Max-Flow Min-Cut Theorem to be described in Section 4) that $f(\lambda, 5) = 4$ is the maximum flow for $T = 5$.

The solution λ is, nevertheless, not an optimum for the universal maximum flow problem. Let us examine it again. When the time limit T is changed from 5 to 4, we have $f(\lambda, 4) = 2$ and λ generates the maximum flow in N within time 4. However, if the time limit is further shortened from 4 to 3, we have $f(\lambda, 3) = 0$. There exists a better solution λ' in this case, which transmits one unit of flow from s at time 0 along the path $P_3 = (s, g, h, \rho)$, which arrives at vertex ρ at time 3.

An optimal solution λ^* that yields the universal maximum flow is to transmit: one unit of flow along the path $P_3 = (s, g, h, \rho)$ at time 0 (which arrives at ρ at 3), one unit of flow along the path $P_1(s, h, \rho)$ at times 0 and 1 respectively (which arrives at ρ at times 4 and 5 respectively), and one unit of flow along the path $P_2 = (s, g, \rho)$ at time 1 (which arrives at ρ at time 5). This gives us $f(\lambda^*, 5) = 4$, $f(\lambda^*, 4) = 2$, $f(\lambda^*, 3) = 1$, $f(\lambda^*, 2) = 0$, and $f(\lambda^*, 1) = 0$.

We now examine the complexity of the TVUMF problem with respect to NP-completeness. It is well-known that the classical maximum flow problem belongs to the class of P. However, the TVUMF problem is NP-complete, as we will show below.

Theorem 3.1 *The TVUMF problem is NP-complete.*

Proof. Obviously, the TVUMF problem is in the class of NP. We now show that the 3-Dimensional matching (3DM) problem can polynomially reduce to the decision version of TVUMF.

The 3DM problem is defined as: Given a set $M \subseteq W \times X \times Y$, where W, X and Y are disjoint sets having the same number q of elements, does M contain a matching, i.e., a subset $M' \subseteq M$ such that $|M'| = q$ and no two elements of M' agree in any coordinate?

The decision version of the TVUMF problem is to answer the question: Given a time-varying network N, a time limit T and an integer k, does there exist a universal flow f from s to ρ within time T such that the value of this flow $v(f) \geq k$?

Figure 3.2. A TVUMF problem constructed from 3DM

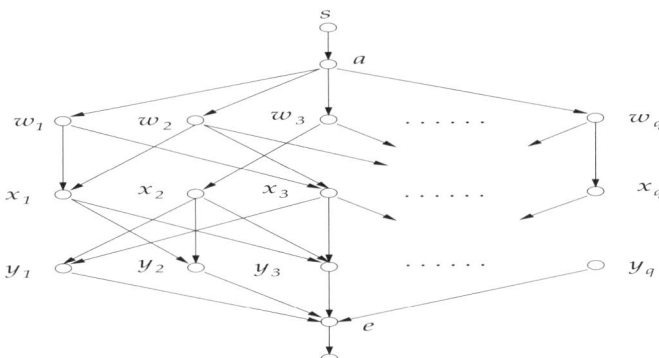

For any given 3DM, we can construct an instance of the TVUMF problem as follows: For each element in W, X and Y, we create vertices w_i, x_i and y_i for $1 \leq i \leq q$. All these vertices, together with s, ρ, and two extra vertices a and e (see Figure 3.2), compose the vertex set V of N. Create arcs (s, a), (e, ρ), (a, w_i) and (y_i, e), $1 \leq i \leq q$, and create arcs (w_i, x_j) and (x_j, y_k) if $(w_i, x_j, y_k) \in M$. All these arcs compose the arc set A of N. The structure of the network is shown in Figure 3.2. Let $l(x, t) = 0$ for each vertex $x \in V$ and for each time $0 \leq t \leq T$. The transit time b and the capacity l_a are defined as follows:

$$b(u, v, t) = 1, \qquad \forall (u, v) \in A, \forall t,$$

$$l(e, \rho, t) = \begin{cases} q, & t = 5, \\ 0, & otherwise, \end{cases}$$

$$l(s, a, t) = \begin{cases} q, & t = 0, \\ 0, & otherwise, \end{cases}$$

$$l(u, v, t) = 1, \qquad \forall (u, v) \in A \backslash \{(e, \rho), (s, a)\}, \forall t.$$

Finally, let $T = 6$ and $k = q$. In what follows, we will prove the claim that the answer to 3DM problem is equivalent to that to the TVUMF problem as constructed above.

Suppose that $M' = \{m_1, m_2, ..., m_q\}$ is a matching of M. For each element $m_i = (w^i, x^i, y^i) \in M'$, we can create a path $P^i = (s, a, w^i, x^i, y^i, e, \rho)$ with departure time 0 from s and arrival time 6 at ρ. Then we can send a subflow, f^i, along P^i with $v(f^i) = 1$, $1 \leq i \leq q$. We can obtain a dynamic flow f, by uniting the q subflows. This is the maximum flow for N, since $v(f) = q$, which is the maximum possible flow value in N.

Given a maximum dynamic flow f in N with $v(f) \geq k = q$, then we must have $v(f) = q$, since q is the maximum possible flow in N. By the structure of N, we must have q disjoint paths, $P_i = (w^i, x^i, y^i)$ $(1 \leq i \leq q)$, from a to e. Let $m_i = (w^i, x^i, y^i)$. By the reduction, we know $m_i \in M$ and none of them agrees in any coordinate. Let $M' = \{m_1, m_2, ..., m_q\}$. It is a matching of M with $|M'| = q$.

This completes the proof. □

From the example discussed above, we can also see that the optimal solution for the TVUMF problem is not unique. In some practical situations, one may want to find the maximum flow with:

1. the earliest departure and the earliest arrival time,
2. the earliest departure and the latest arrival time,
3. the latest departure and the earliest arrival time, or
4. the latest departure and the latest arrival time.

We will discuss these problems in Section 6.

3. The time-varying residual network

Recall that the basic idea of the maximum flow algorithm (Ford et al (1956)) is to find an f-augmenting path from the source vertex to the sink vertex in a residual network and then send as much flow along the path as possible. We will, in this chapter, adopt a similar idea to tackle the TVUMF problem. Nevertheless, because *time* plays an essential role in our model, several concepts will have to be generalized to incorporate those time-varying factors. Specifically, we have to address the following issues:

(1) How to define and find *a dynamic f-augmenting path*, which is feasible in the sense that the flow arrival times and departure times at all its internal vertices are matched.

(2) How to define and generate *a dynamic residual network* after a feasible dynamic f-augmenting path is found and the maximum possible flow along this path is determined.

Based on Definition 1.2, the definition of dynamic path, we now further introduce the concept of the dynamic f-augmenting path.

Definition 3.1 *Let $P(s = x_1, x_2, ..., x_r = x)$ be a dynamic path from s to x. $P(s, x)$ is said to be a dynamic f-augmenting path from s to x if, for $i = 2, ..., r$, it satisfies*

(i) $l(x_{i-1}, x_i, \tau(x_{i-1})) > 0$;

(ii) $\prod_{t=0}^{w(x_{i-1})-1} l(x_{i-1}, \alpha(x_{i-1}) + t) > 0$,

where $0 \le \alpha(x_i) \le T$ and $0 \le \tau(x_i) \le T$ for $i = 1, ..., r$.

Next we generalize the concept of *residual network*. Let us first create a new network from N. For every arc $(x, y) \in A$, create an artificial arc, denoted by $[y, x]$, which has transit time $b[y, x, u]$ and capacity $l[y, x, u]$. For arc $[y, x]$ and $t = 0, 1, ..., T$, let $l[y, x, t] = 0$ initially and define the transit time $b[y, x, t]$ as:

$$b[y, x, t] = \begin{cases} -b(x, y, u) & 0 \le t = u + b(x, y, u) \le T, u = 0, 1, ..., T \\ +\infty & \text{otherwise} \end{cases}$$

Note that $b[y, x, t]$ may take more than one value for some arcs $[y, x]$ at some time t, since there may exist more than one u satisfying $u + b(x, y, u) = t$.

For every vertex $x \in V$, we also define an artificial vertex capacity $l[x, t]$, to represent the capacity under which a flow can "wait" at x from time t to $t - 1$. This definition means that a flow may have a negative waiting time at a vertex x. In fact, similar to the definition of a negative transit time $b[x, y, t]$ that allows us to retract a flow on an arc, the introduction of $l[x, t]$ allows us to retract a waiting time of a flow at vertex x. Initially, let $l[x, t] = 0$ for each x and $t = 1, 2, ..., T$.

Obviously, the new network as created above is equivalent to the original one. Thus we still denote it by N. Let A^+ and A^- denote the non-artificial arc set and the artificial arc set of N, respectively. After induction of artificial arcs, the concept of dynamic f-augmenting path (Definition 3.1) can be further generalized as follows:

Definition 3.2 *Let $P(s = x_1, x_2, ..., x_r = x)$ be a dynamic path from s to x. $P(s, x)$ is said to be a dynamic f-augmenting path from s to x if for $i = 2, ..., r$, it satisfies*

(i) $\tau(x_{i-1}) + b(x_{i-1}, x_i, \tau(x_{i-1})) = \alpha(x_i)$ *if* $(x_{i-1}, x_i) \in A^+$; *or* $\tau(x_{i-1}) + b[x_{i-1}, x_i, \tau(x_{i-1})] = \alpha(x_i)$ *if* $[x_{i-1}, x_i] \in A^-$;

(ii) $l(x_{i-1}, x_i, \tau(x_{i-1})) > 0$, *if* $(x_{i-1}, x_i) \in A^+$; *or* $l[x_{i-1}, x_i, \tau(x_{i-1})] > 0$, *if* $[x_{i-1}, x_i] \in A^-$; *and*

(iii) $\prod_{t=0}^{w(x_{i-1})-1} l(x_{i-1}, \alpha(x_{i-1}) + t) > 0$, *if* $w(x_{i-1}) > 0$; *or* $\prod_{t=0}^{|w(x_{i-1})|-1} l[x_{i-1}, \alpha(x_{i-1}) - t] > 0$, *if* $w(x_{i-1}) < 0$,

where $0 \le \alpha(x_i) \le T$, $0 \le \tau(x_i) \le T$ for $i = 1, ..., r$.

We define the capacity of a dynamic f-augmenting path as follows.

Definition 3.3 *Let $P(s = x_1, x_2, ..., x_r = x)$ be a dynamic f-augmenting path from s to x. The capacity of P is defined as*

$$Cap(P) = \min \Big\{ \min_{(x,y) \in A(P)} l(x, y, \tau(x)), \min_{[x,y] \in A(P)} l[x, y, \tau(x)],$$

$$\min_{v \in V(P), \alpha(x) \leq t' < \tau(x)} l(x, t'), \min_{v \in V(P), \alpha(x) \geq t' > \tau(x)} l[x, t'] \Big\}.$$

Clearly, the capacity of a path is the upper bound of the flow which we can send along the path. In other words, let f_p denote the flow value that can be sent along the path P, we have $f_p \leq Cap(P)$.

After a dynamic f-augmenting path is found, we can send an augmenting flow f_p along it, and then construct a residual network by the following procedure:

Network Updating Procedure-UPNET

Let $P(s, \rho) = (x_1, x_2, ..., x_r)$ be a dynamic f-augmenting path from $s = x_1$ to $\rho = x_r$, and let $f_p > 0$ be the flow value sent along $P(s, \rho)$. For $i = 1, ..., r - 1$, do:

Update arc capacity
    Case I: $(x_i, x_{i+1}) \in A^+$. Let

$$l(x_i, x_{i+1}, \tau(x_i)) := l(x_i, x_{i+1}, \tau(x_i)) - f_p,$$

$$l[x_{i+1}, x_i, \alpha(x_{i+1})] := l[x_{i+1}, x_i, \alpha(x_{i+1})] + f_p$$

    Case II: $[x_i, x_{i+1}] \in A^-$. Let

$$l(x_{i+1}, x_i, \alpha(x_{i+1})) := l(x_{i+1}, x_i, \alpha(x_{i+1})) + f_p,$$

$$l[x_i, x_{i+1}, \tau(x_i)] := l[x_i, x_{i+1}, \tau(x_i)] - f_p$$

For $i = 2, 3, ..., T - 1$ do:

Update vertex capacity
    Case I: $w(x_i) > 0$. Let

$$l(x_i, t) := l(x_i, t) - f_p, \quad t = \alpha(x_i), \alpha(x_i) + 1, ..., \alpha(x_i) + w(x_i) - 1$$

$$l[x_i, t] := l[x_i, t] + f_p, \quad t = \tau(x_i), \tau(x_i) - 1, ..., \tau(x_i) - w(x_i) + 1$$

Case II: $w(x_i) < 0$. Let

$$l[x_i, t] := l[x_i, t] - f_p, \quad t = \alpha(x_i), \alpha(x_i) - 1, ..., \alpha(x_i) + w(x_i) + 1$$

$$l(x_i, t) := l(x_i, t) + f_p, \quad t = \tau(x_i), \tau(x_i) + 1, ..., \tau(x_i) - w(x_i) - 1.$$

Definition 3.4 *The network generated by the procedure above is called a dynamic residual network.*

The problem in the original network and the problem in the dynamic residual network are equivalent in the sense that there is a one-to-one correspondence between their feasible solutions. Note that in the original network we assume that all transit times $b > 0$. Thus, the first dynamic f-augmenting path will only contain arcs with positive transit times b and positive waiting times w. But in a dynamic residual network, the transit time associated with an artificial arc is a negative number, and a flow can be stored at a vertex with a negative waiting time.

Example 3.2

Figure 3.3. The original network of Example 3.2. In the figure, s is the source vertex and ρ is the sink vertex. The three numbers inside each pair of brackets associated with an arc are t, $b(x, y, t)$ and $l(x, y, t)$ respectively. For instance, $(0, 1, 2)$ near the arc (s, d) means at time 0, transit time $b(s, d, 0)$ is 1 and capacity limit $l(s, d, 0)$ is 2, and at other times, $b(s, d, t) = +\infty$ and $l(s, d, t) = 0$. We assume, in this example, that no waiting time at any internal vertex is allowed, so $l(x, t) = 0$ for all x and t.

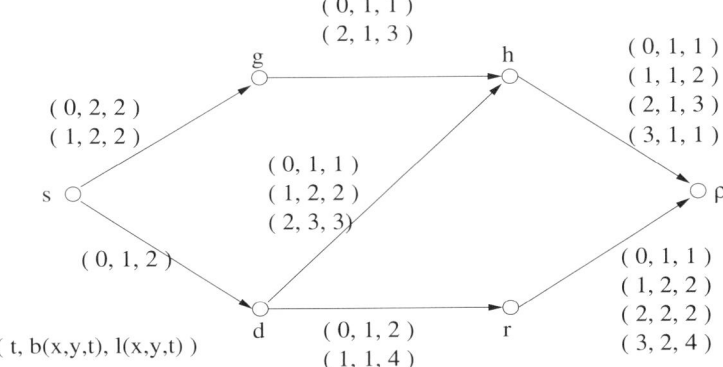

Consider a network N as shown in Figure 3.3, where the three numbers in each bracket associated with an arc are t, $b(x, y, t)$ and $l(x, y, t)$ respectively. For each arc $(x, y) \in A$, we create an artificial arc $[y, x]$; see Figure 3.4, where the two numbers in each bracket on an artificial arc denote t and $b[x, y, t]$, respectively. For other time t that is not shown,

Figure 3.4. The initial network

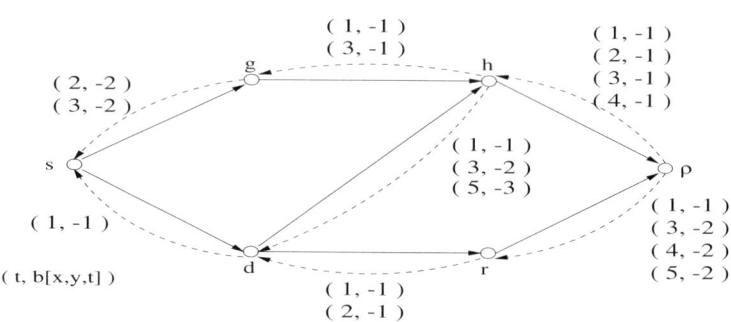

we let $b[x, y, t] = +\infty$. All $l[x, y, t] = 0$. No waiting time at any internal vertex is allowed in this example, and so we let $l(x, t) = 0$ for all x and t.

Figure 3.5. Example 3.2 (continued)

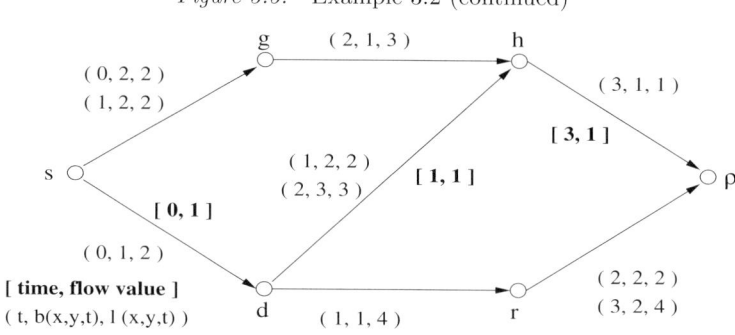

Note that $P_1(s, \rho) = (s, d, h, \rho)$ is a feasible dynamic f-augmenting path from s to ρ. Departing at time 0, at least two units of flow can traverse the arc (s, d), since its capacity at time 0 is 2, and the flow reaches the vertex d at time 1. Similarly, at least two units of flow can be sent through arc (d, h) with departure time 1, and the flow reaches the vertex h at time 3. One unit of flow can depart h at time 3 and arrive at ρ at time 4. Thus, we can send one unit of flow from s to ρ along $P_1(s, \rho)$ within time 4. We label each arc on P_1 with a pair of numbers $[\mathbf{t}, \mathbf{v(f_1)}]$, to denote that a flow value $v(f_1)$ is sent through this arc with departure time t. The resultant network is shown in Figure 3.5 (to simplify the figure, we do not draw the artificial arcs in Figure 3.5).

Then, we update the original network by the Network Updating Procedure. The dynamic residual network created is shown in Figure 3.6.

Figure 3.6. Example 3.2 (continued)

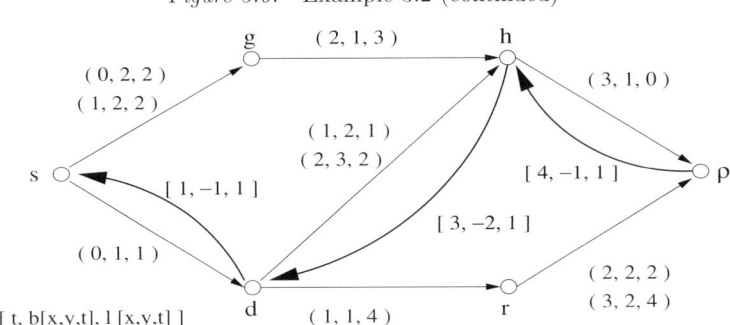

Figure 3.7. Example 3.2 (continued)

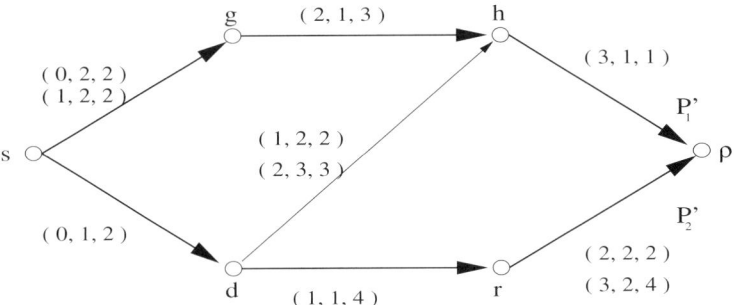

Next, we find another feasible dynamic f-augmenting path $P_2(s, \rho) = (s, g, h, d, r, \rho)$ (note that $[h, d]$ is an artificial arc with negative transit time). At time 0, two units of flow can be sent through (s, g) since $l(s, g, 0) = 2$, which will arrive at g at time 2. At time 2, two units of flow can be sent through (g, h), which will arrive at h at time 3. $[h, d]$ is an artificial arc, from which we can see that one unit of flow was sent in the previous path P_1 during time $[1, 3]$. By travelling from h to d, we actually "push" this one unit of flow back to the vertex d and reduce the time from 3 to 1. And then, since $l(d, r, 1) = 4 > 0$, we can send this one unit of flow through (d, r), which departs at time 1 and reaches r at time 2. Similarly, we can send one unit of flow through (r, ρ), which departs at time 2 and arrives at ρ at time 4. As a result, we can "send" one unit of flow "along" P_2.

Obviously, P_2 is not a real path since it contains an artificial arc, but it helps us determine the real path. In fact, the actual solution is to send the first flow along $P_1' = (s, d, r, \rho)$ and the second flow along $P_2' = (s, g, h, \rho)$. The total flow value that can be sent is $1 + 1 = 2$; see Figure 3.7.

4. The max-flow min-cut theorem

Minimum cut is an important concept in the study of the maximum flow problem. In what follows, we will generalize this definition for the time-varying network and introduce the time-varying version of the max-flow min-cut theorem.

Definition 3.5 *A vertex y is said to be reachable from another vertex x, if there exists a feasible dynamic path from x to y.*

Definition 3.6 *A generalized cut K of N separating vertex s and ρ is a set-valued function of time defined as*

$$K = \{K(t)|K(t) \subset V, s \in K(t), \rho \notin K(t), t = 0, 1, ..., T\}.$$

Definition 3.7 *The capacity of the generalized cut K is defined as*

$$CapK = \sum_{t=0}^{T} \sum_{x \in s(t), y \in \rho(t')} l(x, y, t)$$

$$+ \sum_{t=0}^{T-1} \sum_{(x,u) \in X(t)} \min\{l(x, t), ..., l(x, t + u)\}$$

where $X(t) = \{(x, u)|x \in s(t), x \in h(t + 1), ..., x \in h(t + u), x \in \rho(t + u + 1), 0 \le u < u_x\}$, $s(t) = \{$*all those vertices x in $K(t)$ such that x is reachable from s with arrival time t at x*$\}$, $\rho(t) = \{$*all those vertices x in $\bar{K}(t) = V \backslash K(t)$ such that ρ is reachable from x with departure time t at x*$\}$, $t + b(x, y, t) = t'$, $h(t) = V \backslash \{s(t) \cup \rho(t)\}$, *and* u_x *is a given nonnegative integer.*

Note that we have used u_x to denote the bound of waiting time at vertex x. If $u_x = 0$, then there is no constraint on the waiting time at x. Specifically, when the waiting time at a vertex must be zero or can be arbitrary, Definition 3.7 can be replaced by the following.

Definition 3.8 *The capacity of the generalized cut can be defined as*

$$CapK = \sum_{t=0}^{T} \sum_{x \in K(t), y \in \bar{K}(t')} l(x, y, t)$$

if no waiting time is allowed at any vertices, or

$$CapK = \sum_{t=0}^{T} \sum_{x \in K(t), y \in \bar{K}(t')} l(x, y, t) + \sum_{t=0}^{T-1} \sum_{x \in K(t), x \in \bar{K}(t+1)} l(x, t)$$

if waiting time is arbitrarily allowed at any vertices, where $t' = t + b(x, y, t)$.

Theorem 3.2 *Let* v *be the value of any feasible dynamic flow* f *on* N, *and* $CapK$ *be the value of any generalized cut* K *in* N. *Then,* $v \leq CapK$.

Proof. Since any generalized cut K separates s and ρ for any dynamic flow f, we have

$$v = f^+(K) - f^-(K)$$

where $f^+(K)$ and $f^-(K)$ are the flow values that flow out of and flow in K respectively. Because f is a dynamic flow on N, it must satisfy all capacity constraints. Therefore

$$0 \leq f(x, y, t) \leq l(x, y, t), \quad \forall (x, y) \in A \text{ and } \forall t,$$

$$0 \leq f(x, t) \leq l(x, t), \quad \forall x \in V \text{ and } \forall t.$$

Thus, we have

$$f^+(K) = \sum_{t=0}^{T} \sum_{x \in K(t), y \in \bar{K}(t+b(x,y,t))} f(x, y, t) + \sum_{t=0}^{T-1} \sum_{x \in K(t), x \in \bar{K}(t+1))} f(x, t)$$

$$= \sum_{t=0}^{T} \sum_{x \in s(t), y \in \rho(t')} f(x, y, t) + \sum_{t=0}^{T-1} \sum_{(x,u) \in X(t)} f(x, u)$$

$$\leq \sum_{t=0}^{T} \sum_{x \in s(t), y \in \rho(t')} l(x, y, t) + \sum_{t=0}^{T-1} \sum_{(x,u) \in X(t)} \min\{l(x, t), ..., l(x, t + u)\}$$

$$= CapK$$

The second equality comes from the fact that $f(x, y, t) = 0$ for those vertices $x \in K(t) \backslash s(t)$, since there is no dynamic path from s to x at time t. Noting that $f^-(K) \geq 0$, we have

$$v = f^+(K) - f^-(K) \leq CapK$$

This completes the proof. $\square$

5. A condition on the feasibility of f-augmenting paths

As we have mentioned have, f-augmenting path plays an important role in solving the maximum flow problem. A related question we have to explore is, in a dynamic residual network, how to determine whether a dynamic f-augmenting path is feasible if there is a waiting constraint at a vertex. Note that by saying a dynamic f-augmenting path is feasible, we mean that we can send an augmenting flow on this path such that the resulting actual flow (combining the previous flow with the current augmenting flow) is feasible under all constraints, including the constraints on the waiting time at vertices. Let us first consider an example to examine the waiting time constraint.

Example 3.3

Figure 3.8.   How to determine the departure time at a vertex such that the resulted f-augmenting path is feasible under the bounded waiting time constraint

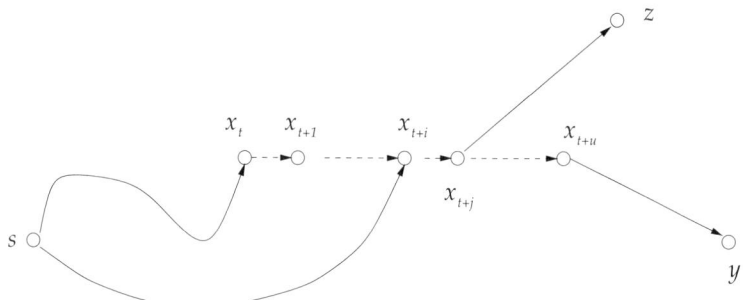

Suppose that we have a dynamic path P from s to x with $\alpha(x) = t$, and we want to append an arc (x, y) to P (see Figure 3.8, where we split the vertex x into x_t, x_{t+1}, ..., x_{t+u} to represent the states of x at different times). Let u_x be the upper bound of the waiting time at x. A problem is to find the latest possible departure time at x, denoted by $t + u$, such that the new path P' from s to y, obtained by adding arc (x, y) to P, is still feasible in terms of satisfying the upper bound u_x. Clearly, if $u \leq u_x$, then P' is feasible since the waiting time at x is not greater than the upper bound u_x. But an observation shows that u could be greater than u_x. Let us look at the following scenario.

Suppose that there is another subflow, denoted by f', in N already, which also passes through vertex x, with the arrival time $t + i$ and the departure time time $t + j$ ($i < j < u_x$) (see Figure 3.8). In this case, the latest departure time at x on path P can be $t + i + u_x$. In other words, for path P, any u satisfying $i + u \leq u_x$ can be chosen as a departure time at x on P, and the new path P is still feasible. In fact, we can reconstruct two paths, P and $\bar{P}$, as follows: appending arc (x, z) to P with the departure time $t + j$ to obtain path P_1, and appending arc (x, y) to path $\bar{P}$ with the departure time $t + i + u_x$. Obviously, both of the two paths satisfy the bounded waiting time constraint; see Figure 3.9.

Figure 3.9. The two paths P and $\bar{P}$

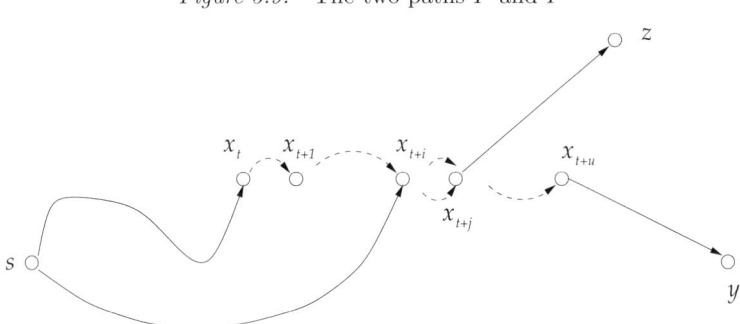

The example above tells us that, for a dynamic f-augmenting path, the waiting time at a vertex x could be greater than u_x while the actual waiting constraint is still satisfied. In what follows, we will give a feasible waiting time condition for a dynamic f-augmenting path under the constraint that the waiting time at a vertex x is bounded above by u_x. This condition is also applicable to the cases with $u_x = 0$ and $u_x = \infty$.

Without loss of generality, assume that $P(s, x)$ is a feasible dynamic f-augmenting path from s to x; (z, x) (or $[z, x]$) is its last arc; and $\alpha(x) = t$ is its arrival time at the vertex x. We now consider how to select the vertex y, or equivalently, to append an arc (x, y) (or $[x, y]$) to $P(s, x)$ (denote this new path as $P(s, x, y)$), with a suitable departure time $\tau(x)$ at x, so as to make $P(s, x, y)$ still feasible. For this purpose, we first examine the following cases, and then describe the feasibility condition.

Case I. Both (z, x) and (x, y) are not artificial arcs. Use $\bar{f}$ to denote the previous flow from which the present dynamic residual network is created. Decompose $\bar{f}$ into subflows flows $f_1, f_2, ..., f_k$ in such a way that f_i traverses the path $P_i(s, \rho)$. Moreover, for f_i and f_j with $i < j$, if they pass through a vertex x, then $\alpha_i(x) \le \alpha_j(x)$ and $\tau_i(x) \le \tau_j(x)$, where $\alpha_i(x), \alpha_j(x)$ and $\tau_i(x), \tau_j(x)$ are arrival times and departure times of f_i and f_j at vertex x, respectively. We may show that in this case, $P(s, x, y)$ *is feasible when*

(i) $\tau(x) \in [\alpha(x), \alpha(x) + u_x]$ *and* $\prod_{t'=\alpha(x)}^{\tau(x)-1} l(x, t') > 0$, *if there does not exist* $1 \le i \le k$ *such that* $[\alpha_i(x), \tau_i(x)] \subseteq [\alpha(x), \alpha(x) + u_x]$, *or*

(ii) $\tau(x) \in [\alpha(x), \alpha_{i^0}(x) + u_x]$ *and* $\prod_{t'=\alpha(x)}^{\tau(x)-1} l(x, t') > 0$, *if for* $1 \le i_1 < i_2 < ... < i^0 \le k$, *there exist* $[\alpha_{i_1}(x), \tau_{i_1}(x)] \subseteq [\alpha(x), \alpha(x) + u_x]$, $[\alpha_{i_2}(x), \tau_{i_2}(x)] \subseteq [\alpha_{i_1}(x), \alpha_{i_1}(x) + u_x]$, ..., $[\alpha_{i^0}(x), \tau_{i^0}(x)] \subseteq [\alpha_{i^0-1}(x), \alpha_{i^0-1}(x) + u_x]$, *or*

(iii) $\tau(x) \in [0, \alpha(x) - 1]$ *and* $\prod_{t'=0}^{\alpha(x)-\tau(x)-1} l[x, \alpha(x) - t'] > 0$.

Condition (i) above is straightforward, since if $\tau(x) \in [\alpha(x), \alpha(x)+u_x]$, we have $w(x) = \tau(x) - \alpha(x) \leq u_x$, and thus $P(s,x,y)$ is feasible. For condition (ii), we have $\tau_{i_1}(x)-\alpha(x) \leq u_x$, $\tau_{i_2}(x)-\alpha_{i_1}(x) \leq u_x$,...,$\tau_{i^0}(x)-\alpha_{i^0-1}(x) \leq u_x$. Thus, we can construct new paths $P'_{i'}$ ($i' = i_1, ..., i^0$) by the following method: let $P'_{i_1} = P(s,x) \cup P_{i^0}(x,\rho)$, $P'_{i_2} = P_{i_1}(s,x) \cup P_{i_2}(x,\rho)$,..., $P'_{i^0} = P_{i^0_1}(s,x) \cup P_{i^0}(x,\rho)$ and $P'_{i^0+1} = P_{i^0}(s,x) \cup (x,y)$. Denote the capacity of the path $P(s,x,y)$ as $Cap(P(s,x,y))$ and let $\delta = \min\{f_{i_1}, ..., f_{i^0}, Cap(P(s,x,y))\}$. And then, decrease $f_{i'}$ to $f_{i'} - \delta$ and send the difference δ on path $P'_{i'}$ ($i' = i_1, ..., i^0$). Again, send δ units of augmenting flow on path P'_{i^0+1}. Since $w'_{i'}(x) \leq u_x$, and other waiting times are unchanged, all paths $P'_{i'}$ are feasible. This means $P(s,x,y)$ is feasible too (see Figure 3.10, where the scenarios (a) and (b) correspond to conditions (ii) and (iii) above, respectively. To simplify the picture, we only draw one subflow flow f_1 in the graph).

Figure 3.10.   Splitting vertex x into several dummy vertices to represent its states at different times.

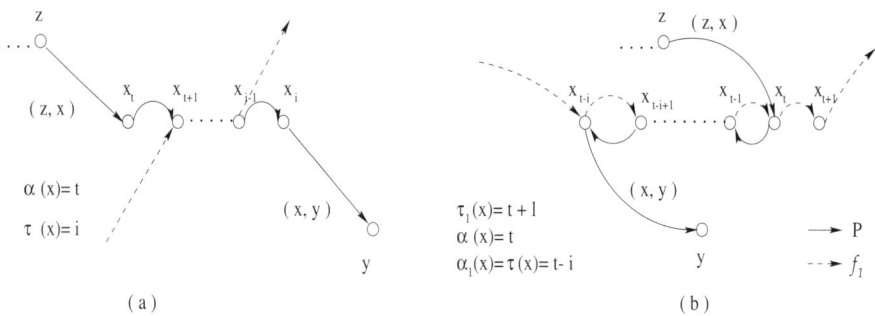

We now justify condition (iii). Without loss of generality, suppose that $f_1, f_2, ..., f_h$ ($h \leq k$) pass through the vertex x. Since $l[x,t] > 0$ for $t = \tau(x)+1, ..., \alpha(x)-1, \alpha(x)$, we have $[\tau(x), \alpha(x)] \subseteq \cap_{i=1}^{h}[\alpha_i(x), \tau_i(x)]$. Note that we have $\tau_h(x) \geq \alpha(x)$, $\alpha_1(x) \leq \tau(x)$ and $\tau_h(x) - \alpha(x) \leq u_x$, $\tau(x) - \alpha_1(x) \leq u_x$. Similar to (ii), we can construct $P'_1, ..., P'_h$ and $P'(s,x,y)$ by the following method: let $P'_h = P(s,x) \cup P_h(x,\rho)$, $P'_i = P_{i+1}(s,x) \cup P_i(x,\rho)$ ($i = 1, ..., h-1$) and $P'_{h+1} = P_1(s,x) \cup (x,y)$. Then, let $\delta = \min\{f_1, ..., f_h, Cap(P(s,x,y))\}$, decrease f_i to $f_i - \delta$ and send the difference δ on path P'_i ($i = 1, ..., h$). Again, send δ units of augmenting flow on path P'_{h+1}. Since $w'_i(x) \leq u_x$ for $i = 1, ..., h+1$, and other waiting times are unchanged, P'_i, $i = 1, ..., h+1$, are feasible. This means $P(s,x,y)$ is also feasible.

Case II. (z,x) **is not an artificial arc and** $[x,y]$ **is an artificial one**. Similar to Case I, we decompose the previous flow $\bar{f}$ into some subflows f_i ($i = 1, ..., k$). Since $l[x, y, \tau(x)] > 0$, some of f_i may traverse arc (y, x) with a common arrival time. Without loss of generality, suppose these subflows are $f_1, f_2, ..., f_{h_1}$ ($h_1 \leq k$). Let $\alpha_i(x)$ and $w_i(x)$ be their arrival times and waiting times at x ($i = 1, ..., h_1$). We have $\alpha_i(x) = \tau(x)$. Letting $w^0(x) = \min_i w_i(x)$, we may show that $P(s, x, y)$ *is feasible when*

(i) $\tau(x) \in [\alpha(x), \alpha(x) - w^0(x) + u_x]$ *and* $\prod_{t'=\alpha(x)}^{\tau(x)-1} l(x, t') > 0$, *if there does not exist* $1 \leq i \leq k$ *such that* $[\alpha_i(x), \tau_i(x)] \subseteq [\alpha(x), \alpha(x) + u_x]$, *or*

(ii) $\tau(x) \in [\alpha(x), \alpha_{i^0}(x) - w^0(x) + u_x]$ *and* $\prod_{t'=\alpha(x)}^{\tau(x)-1} l(x, t') > 0$, *if for* $1 \leq i_1 < i_2 < ... < i^0 \leq k$, *there exist* $[\alpha_{i_1}(x), \tau_{i_1}(x)] \subseteq [\alpha(x), \alpha(x) + u_x]$, $[\alpha_{i_2}(x), \tau_{i_2}(x)] \subseteq [\alpha_{i_1}(x), \alpha_{i_1}(x) + u_x], ..., [\alpha_{i^0}(x), \tau_{i^0}(x)] \subseteq [\alpha_{i^0-1}(x), \alpha_{i^0-1}(x) + u_x]$, *or*

(iii) $\tau(x) \in [0, \alpha(x) - 1]$ *and* $\prod_{t'=0}^{\alpha(x)-\tau(x)-1} l[x, \alpha(x) - t'] > 0$.

Denote $P^0(s, \rho)$ as the feasible dynamic path corresponding to $w^0(x)$. Note that $[x, y]$ is an artificial arc. So $P(s, x, y)$ is not a true path. However, we can construct a new path $P' = P(s, x) \cup P^0(x, \rho)$. Let $w'(x)$ be the waiting time of P' at x. Clearly, if $\tau(x) \in [\alpha(x), \alpha(x) - w^0(x) + u_x]$, then $w'(x) = \tau(x) - \alpha(x) + w^0(x) \leq u_x$. This together with the satisfaction of the capacity constraint means that P' is feasible and justifies condition (i) above. The justification of conditions (ii) and (iii) is similar to that in Case I (also see Figure 3.11(b) below).

Figure 3.11. The feasibility condition for Case II

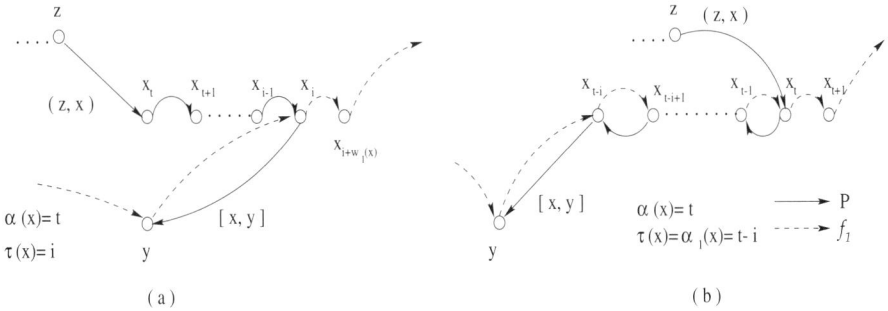

(a)   (b)

Case III. $[z, x]$ is an artificial arc and (x, y) is not. Since $l[z, x, \tau(z)] > 0$, similar to Case II, we know that some of f_i traverse the arc (x, z) with a common departure time. Without loss of generality, suppose these subflows are $f_1, f_2, ..., f_{h_2}$ ($h_2 \leq k$). Let $\alpha_i(x)$ and $w_i(x)$ be their arrival times and waiting times at x ($i = 1, ..., h_2$). We have $\alpha_i(x) = \alpha(x)$. Letting $w^*(x) = \min_i w_i(x)$, we may show that

$P(s, x, y)$ *is feasible when*

(i) $\tau(x) \in [\alpha(x), \alpha(x) - w^*(x) + u_x]$ *and* $\prod_{t'=\alpha(x)}^{\tau(x)-1} l(x, t') > 0$, *if there does not exist* $1 \leq i \leq k$ *such that* $[\alpha_i(x), \tau_i(x)] \subseteq [\alpha(x), \alpha(x) + u_x]$, *or*

(ii) $\tau(x) \in [\alpha(x), \alpha_{i^0}(x) - w^*(x) + u_x]$ *and* $\prod_{t'=\alpha(x)}^{\tau(x)-1} l(x, t') > 0$, *if for* $1 \leq i_1 < i_2 < ... < i^0 \leq k$, *there exist* $[\alpha_{i_1}(x), \tau_{i_1}(x)] \subseteq [\alpha(x), \alpha(x) + u_x]$, $[\alpha_{i_2}(x), \tau_{i_2}(x)] \subseteq [\alpha_{i_1}(x), \alpha_{i_1}(x) + u_x], ..., [\alpha_{i^0}(x), \tau_{i^0}(x)] \subseteq [\alpha_{i^0-1}(x), \alpha_{i^0-1}(x) + u_x]$, *or*

(iii) $\tau(x) \in [0, \alpha(x) - 1]$ *and* $\prod_{t'=0}^{\alpha(x)-\tau(x)-1} l[x, \alpha(x) - t'] > 0$.

When $\tau(x) \in [\alpha(x), \alpha(x) - w^*(x) + u_x]$, we can construct a new path $P' = P^*(s, x) \cup (x, y)$, where P^* is the dynamic path corresponding to $w^*(x)$. The waiting time of P' at x, denoted by $w'(x)$, will be $w'(x) = w^*(x) + \tau(x) - \alpha(x) \leq u_x$, and thus P' is feasible under condition (i) above. As for conditions (ii) and (iii), the discussion is similar to that in Case I.

Case IV. Both $[z, x]$ **and** $[x, y]$ **are artificial arcs.** Similar to Case I, decompose $\bar{f}$ to subflows f_i ($i = 1, ..., k$). Since $l[z, x, \tau(z)] > 0$ and $l[x, y, \tau(x)] > 0$, some of subflows may traverse arc (x, z) or (y, x) with a common departure time or a common arrival time. Let $f_i^{(1)}$ ($i = 1, ..., h$) denote those subflows that traverse the arc (x, z) with a common departure time $\tau_i^{(1)}(x) = \alpha(x)$, and let $f_j^{(2)}$ ($j = 1, ..., e$) denote those subflows that traverse the arc (y, x) with a common arrival time $\alpha_j^{(2)}(x) = \tau(x)$. Let $w^{(1)}(x) = \min_i w_i^{(1)}(x)$ and $w^{(2)}(x) = \min_j w_j^{(2)}(x)$. Then one may show that $P(s, x, y)$ *is feasible if*

(i) $\tau(x) \in [\alpha(x), \alpha(x) - w^{(1)}(x) - w^{(2)}(x) + u_x]$ *and* $\prod_{t'=\alpha(x)}^{\tau(x)-1} l(x, t') > 0$, *if there does not exist* $1 \leq i \leq k$ *such that* $[\alpha_i(x), \tau_i(x)] \subseteq [\alpha(x), \alpha(x) + u_x]$, *or*

(ii) $\tau(x) \in [\alpha(x), \alpha_{i^0}(x) - w^{(1)}(x) - w^{(2)}(x) + u_x]$ *and* $\prod_{t'=\alpha(x)}^{\tau(x)-1} l(x, t') > 0$, *if for* $1 \leq i_1 < i_2 < ... < i^0 \leq k$, *there exist* $[\alpha_{i_1}(x), \tau_{i_1}(x)] \subseteq [\alpha(x), \alpha(x) + u_x]$, $[\alpha_{i_2}(x), \tau_{i_2}(x)] \subseteq [\alpha_{i_1}(x), \alpha_{i_1}(x) + u_x], ..., [\alpha_{i^0}(x), \tau_{i^0}(x)] \subseteq [\alpha_{i^0-1}(x), \alpha_{i^0-1}(x) + u_x]$, *or*

(iii) $\tau(x) \in [0, \alpha(x) - 1]$ *and* $\prod_{t'=0}^{\alpha(x)-\tau(x)-1} l[x, \alpha(x) - t'] > 0$.

The case above is a combination of Case II with Case III. We thus omit the detailed justification here.

From the discussion above, we see that when an artificial arc is to be appended to path P, the waiting times of the previous flow will have to be considered. We need to modify the Network Update Procedure (see Section 2) to record this information when updating arc capacities.

First, for any artificial arc $[x, y]$ and any time t, let $r_x^j[x, y, t]$ and $r_y^j[x, y, t]$ denote the waiting times of the previous flow f_j ($j = 1, ..., h$), which traverse the arc (y, x) with the arrival time t, at x and y respectively. Let $g_x^j[x, y, t]$ and $g_y^j[x, y, t]$ denote the values of f_j correspondingly. Furthermore, let us set up two queues, $R_x[x, y, t]$ and $R_y[x, y, t]$, to contain the binary elements $(r_x^j[x, y, t], g_x^j[x, y, t])$ and $(r_y^j[x, y, t], g_y^j[x, y, t])$, respectively. Initially, let $R_x := R_y := \emptyset$. Since we always need the minimal waiting times (see the discussion above), we maintain R_x and R_y in nondecreasing order for r_x^j and r_y^j respectively (in the rest of this paper, we assume that r_x^1 and r_y^1 are the minimal ones). Suppose $P(s = x_1, x_2, ..., x_r = \rho)$ is a feasible dynamic f-augmenting path we obtained and consider the following two cases:

(i) $(x_i, x_{i+1}) \in A^+$. We let in the Network Update Procedure,

$$r_{x_i}[x_{i+1}, x_i, \alpha(x_{i+1}))] := w(x_i), \qquad r_{x_{i+1}}[x_{i+1}, x_i, \alpha(x_{i+1})] := w(x_{i+1})$$

$$g_{x_i}[x_{i+1}, x_i, \alpha(x_{i+1})] := g_{x_{i+1}}[x_{i+1}, x_i, \alpha(x_{i+1})] := f_p$$

and insert (r_{x_i}, g_{x_i}) and $(r_{x_{i+1}}, g_{x_{i+1}})$ in R_{x_i} and $R_{x_{i+1}}$ respectively, while keeping r_{x_i} and $r_{x_{i+1}}$ in nondecreasing order.

(ii) $[x_i, x_{i+1}] \in A^-$. We decrease the values of g_{x_i} and $g_{x_{i+1}}$ accordingly while decreasing the value of $l[x_i, x_{i+1}, \tau(x_i)]$. If g_{x_i} (or $g_{x_{i+1}}$) becomes zero, delete (r_{x_i}, g_{x_i}) from R_{x_i} (or $(r_{x_{i+1}}, g_{x_{i+1}})$ from $R_{x_{i+1}}$).

Second, for each vertex $x \in V \backslash \{s, \rho\}$, let $\alpha_i(x, t)$ denote the arrival time of the subflow f_i at x which departs from x at time t. If no subflow passes x or a subflow passes x but it does not depart at time t, then set $\alpha_1(x, t) = -\infty$. Furthermore, set up a queue $H(x, t)$ to contain $\alpha_i(x, t)$. Initially, $H(x, t) = \emptyset$. We maintain H in nondecreasing order of $\alpha_i(x, t)$. Suppose P is a feasible dynamic f-augmenting path obtained. We can now find $\alpha_{i^0}(x)$ as follows:

Let

$$\zeta_1 = \max_{t_1 = \alpha(x)+1, ..., \alpha(x)+u_x, t_1 \leq T} \alpha_1(x, t_1)$$

If $\zeta_1 > \alpha(x)$, then we say that $\alpha_0(x)$ *does not exist*. Otherwise, let

$$\zeta_2 = \max_{t_2 = \zeta_1+1, ..., \zeta_1+u_x, t_2 \leq T} \alpha_1(x, t_2)$$

Keep on doing the above computation while $\zeta_i + u_x \leq T$, which will give us a series $\zeta_1 < \zeta_2 < ... < \zeta_{i^0}$. Then, $\alpha_{i^0}(x) = \zeta_{i^0}$.

In summary, we have

Theorem 3.4 *Let $P(s = x_1, ..., x_r = \rho)$ be a dynamic f-augmenting path with all $\alpha(x_i)$, $w(x_i)$ and $\tau(x_i)$ ($t = 1, 2, ..., T$) in a dynamic residual network, and P satisfies capacity constraints both on arcs and at*

vertices. P is feasible under the bounded waiting time constraint iff for
$i = 2, 3, ..., r - 1,$

(i) $\tau(x_i) \in [0, \alpha(x_i) - 1]$ *and* $\prod_{t'=0}^{\alpha(x_i)-\tau(x_i)-1} l[x_i, \alpha(x_i) - t'] > 0;$ *or*

(ii) $\tau(x_i) \in [\alpha(x_i), \alpha_{i^0}(x_i) + u_{x_i}],$ *where* $\alpha_{i^0}(x_i)$ *is defined as in Case I, or*

(a) $\tau(x_i) \in [\alpha(x_i), \alpha(x_i) + u_{x_i}],$ *if both* (x_{i-1}, x_i) *and* (x_i, x_{i+1}) *are not artificial and there does not exist* $1 \leq j \leq k$ *such that* $[\alpha_j(x_i), \tau_j(x_i)] \subseteq [\alpha(x_i), \alpha(x_i) + u_{x_i}];$

(b) $\tau(x_i) \in [\alpha(x_i), \alpha(x_i) - r_{x_i}^1[x_i, x_{i+1}, \tau(x_i)] + u_{x_i}],$ *if* (x_{i-1}, x_i) *is not artificial but* $[x_i, x_{i+1}]$ *is artificial and there does not exist* $1 \leq j \leq k$ *such that* $[\alpha_j(x_i), \tau_j(x_i)] \subseteq [\alpha(x_i) - r_{x_i}^1[x_i, x_{i+1}, \tau(x_i)], \alpha(x_i) - r_{x_i}^1[x_i, x_{i+1}, \tau(x_i)] + u_{x_i}];$

(c) $\tau(x_i) \in [\alpha(x_i), \alpha(x_i) - r_{x_i}^1[x_{i-1}, x_i, \tau(x_{i-1})] + u_{x_i}],$ *if* $[x_{i-1}, x_i]$ *is artificial but* (x_i, x_{i+1}) *is not and there does not exist* $1 \leq j \leq k$ *such that* $[\alpha_j(x_i), \tau_j(x_i)] \subseteq [\alpha(x_i) - r_{x_i}^1[x_{i-1}, x_i, \tau(x_{i-1})], \alpha(x_i) - r_{x_i}^1[x_{i-1}, x_i, \tau(x_{i-1})] + u_{x_i}];$

(d) $\tau(x_i) \in [\alpha(x_i), \alpha(x_i) - r_{x_i}^1[x_{i-1}, x_i, \tau(x_{i-1})] - r_{x_i}^1[x_i, x_{i+1}, \tau(x_i)] + u_{x_i}],$ *if both* $[x_{i-1}, x_i]$ *and* $[x_i, x_{i+1}]$ *are artificial and there does not exist* $1 \leq j \leq k$ *such that* $[\alpha_j(x_i), \tau_j(x_i)] \subseteq [\alpha(x_i) - r_{x_i}^1[x_i, x_{i+1}, \tau(x_i)] - r_{x_i}^1[x_{i-1}, x_i, \tau(x_{i-1})], \alpha(x_i) - r_{x_i}^1[x_i, x_{i+1}, \tau(x_i)] - r_{x_i}^1[x_{i-1}, x_i, \tau(x_{i-1})] + u_{x_i}].$

Proof. From the justification in the four cases above, it is clear that these conditions are sufficient. We now show that they are also necessary.

Assume that $P(s = x_1, x_2, ..., x_r = \rho)$ is a feasible dynamic f-augmenting path with a departure time $\tau(x_i)$ that does not satisfy the conditions. We consider the case in which both (x_{i-1}, x_i) and (x_i, x_{i+1}) are not artificial.

(i) If $\tau(x_i) \in [0, \alpha(x_i) - 1],$ then $\prod_{t'=0}^{\alpha(x_i)-\tau(x_i)-1} l[x_i, \alpha(x_i) - t'] = 0,$ since $\tau(x_i)$ does not satisfy the conditions. By Definition 4, P is not a feasible dynamic f-augmenting path, even if its waiting time is arbitrary. Clearly, this is a contradiction.

(ii) Suppose that $\tau(x_i) \in [\alpha(x_i) + u_{x_i} + 1, +\infty].$ One can see that there must exist at least one previous flow waiting at x_i during the period $[\alpha_j(x_i), \tau_j(x_i)] \subseteq [\alpha(x_i), \alpha(x_i) + u_{x_i}],$ since if such a flow does not exist, then the waiting time of P at x_i, $w(x_i) = \tau(x_i) - \alpha(x_i) \geq u_{x_i} + 1.$ We can create new paths to shorten the waiting time $w(x_i)$ by changing the flowing directions (see the method we described in Case I (ii)). However, to shorten $w(x_i)$, this previous flow, $f_{j'}$, must have $\alpha(x_i) \leq \alpha_{j'}(x_i) \leq \alpha(x_i) + u_{x_i}$ and $\tau_{j'}(x_i) \leq \alpha(x_i) + u_{x_i},$ i.e., $[\alpha_{j'}(x_i), \tau_{j'}(x_i)] \subseteq [\alpha(x_i), \alpha(x_i) + u_{x_i}].$ This is also a contradiction.

(iii) Suppose that $\tau(x_i) \in [\alpha_{i^0}(x_i) + u_{x_i} + 1, +\infty]$ and there does not exist a previous flow, f_j, such that $[\alpha_j(x_i), \tau_j(x_i)] \subseteq [\alpha_{i^0}(x_i), \alpha_{i^0}(x_i) + u_{x_i}]$. This situation is similar to (ii) if we replace $\alpha_{i^0}(x_i) + u_{x_i}$ by $\alpha(x_i) + u_{x_i}$, and thus we omit the detailed proof.

The proof in other cases can also be conducted in a similar way. □

6. Algorithms

In this section, we will describe a few label setting algorithms, which take into consideration the structure of waiting constraints. We first deal with the case where no waiting times are permitted at any vertices, i.e., $u_x = 0$ for all $x \in V$.

Since no waiting time is permitted at any vertex, we need not consider the capacity of a vertex (see our definition of vertex capacity in Section 2). The algorithm to be constructed is based on an observation that, at each time $t' \in [0, t]$ for any given t, if we augment as much flow as possible from the source to the sink in the dynamic residual network, then the total value of the flows augmented within the period $[0, t]$ will consist of the maximum flow for that period. As the maximum flow for each $[0, t]$ can be obtained as t grows from 0 to T, such a procedure gives naturally the *universal* solution we aim to achieve. One may see better the validity of the above observation from the proof of Theorem 4.4 in the previous section. More specifically, the algorithm will perform the following operations:

(i) Maintain a set of labels $d(y, 0)$, $d(y, 1)$,..., $d(y, T)$ for each vertex y, where $d(y, t)$ is set to be the predecessor of y in a feasible dynamic f-augmenting path $P(s, y)$ of time exactly t if this path exists (for source vertex s, define its predecessor to be 0); otherwise, if such a path does not exist, set $d(y, t) = null$. A vertex y is said to be "labelled" at time t if $d(y, t) \neq null$.

(ii) Initially, set all $d(s, t)$ $(t = 0, 1, ..., T)$ to be 0, and construct a queue, Q, to contain them. Then, while Q is not empty, perform a *labelling operation*: choose the first label $d(x, t)$ from Q, and for each y with $(x, y) \in A^+$ (or $[x, y] \in A^-$) and $t' = t + b(x, y, t)$ (or each $t' = t + b[x, y, t]$), set $d(y, t') = \{x\}$ if the following conditions hold:

(1) $0 \leq t' \leq T$;

(2) $d(y, t') = null$;

(3) $l(x, y, t) > 0$ (or $l[x, y, t] > 0$).

Delete $d(x, t)$ from Q and append $d(y, t')$, if $d(y, t') = \{x\}$ (that is, the conditions above hold), to Q.

(iii) Check all labels $d(\rho, t)$ $(t = 1, 2, ..., T)$. If there is a label $d(\rho, t) \neq null$ $(t = 0, 1, ..., T)$, then there is a feasible dynamic f-augmenting path from s to ρ of time exactly t. Choose the one with the earliest arriving time t if there are multiple labels which are not null. Then send the maximum possible flow along this path, construct the dynamic residual network accordingly, reset all labels $d(y, t)$ to null and try to find another path; otherwise, stop the algorithm, and the sum of the flows found during the process gives the final solution.

We now describe the details of the algorithm, where the notation $g(t_{min})$ denotes the augmenting flow value along the feasible dynamic f-augmenting path $P(s, \rho)$ of time exactly t_{min} found in one iteration, and $maxf$ denotes the maximum flow value.

> **Algorithm TVUMF-ZW**
> **Begin** $maxf := 0$; $t_{min} := 0$;
> **While** $t_{min} \leq T$ **do**
> Initialization: $d(s, t) := 0, t = 0, 1, ..., T$;
> $\qquad\qquad Q := \{d(s, 0), d(s, 1), ..., d(s, T)\}$;
>   **While** $Q \neq \emptyset$ **do**
>   Select the first label in Q, denoted by $d(x, t)$;
>   **For** any y such that (x, y) or $[x, y] \in A$ **do**
>   **Let** $t' := t + b(x, y, t)$ (or $t' := t + b[x, y, t]$ for each value of $b[x, y, t]$);
>   **If** $t < t' \leq T$ **then** $c' := l(x, y, t)$ **Else** $c' := l[x, y, t]$;
>   *labelling:*
>   **If** $(0 \leq t' \leq T)$And$(d(y, t') = null)$And$(c' > 0)$ **then** $d(y, t') := x$; $Q := Q \cup \{d(y, t')\}$;
>   **Let** $Q := Q \backslash \{d(x, t)\}$;
>   Let t_{min} be the minimum t such that $d(\rho, t) \neq null$; If $d(\rho, t) = null$ for all t, then let $t_{min} = T + 1$ (the algorithm will terminate);
>   **If** $t_{min} < T + 1$ **then**
>   Use the predecessor indices to identify the feasible dynamic f-augmenting path $P(s, \rho)$ of time exactly t_{min};
>   **Let** $g(t_{min}) :=$ the minimum capacity of arcs in $P(s, \rho)$;
>   Augment $g(t_{min})$ units of flow along $P(s, \rho)$;
>   Call procedure UPNET to update the arc and vertex capacities;
>   **Let** $maxf := maxf + g(t_{min})$;
>   **End**.

The following example illustrates how the algorithm works.

Example 3.4

In Figure 3.12, the solid lines stand for the original arcs of network N, and the dotted lines denote the artificial arcs. Initially, for all artificial

Table 3.1. $b(x, y, t)$ and $l(x, y, t)$

t	(s, q)	(s, g)	(s, z)	(q, v)	(v, w)	(g, h)
0	1, 1	1, 3	1, 2	1, 1	1, 3	1, 3
1	1, 0	3, 0	1, 0	1, 2	1, 3	2, 2
2	1, 0	3, 0	2, 0	1, 1	1, 2	2, 0
3	3, 0	1, 0	1, 0	2, 3	2, 3	1, 0
4	2, 0	1, 0	3, 0	1, 2	1, 2	2, 2
5	1, 0	2, 0	4, 0	3, 1	2, 2	1, 0
6	1, 0	1, 0	1, 0	3, 2	1, 1	2, 0
7	2, 0	2, 0	1, 0	2, 1	3, 3	1, 0
8	-, -	-, -	-, -	-, -	-, -	-, -

t	(g, r)	(z, r)	(w, g)	(w, ρ)	(h, ρ)	(r, ρ)
0	1, 2	1, 3	1, 4	1, 3	1, 3	1, 2
1	1, 2	1, 2	1, 2	3, 3	1, 4	1, 2
2	2, 0	1, 0	1, 2	3, 2	1, 0	1, 2
3	2, 0	2, 0	1, 1	2, 0	1, 4	5, 1
4	2, 0	2, 0	3, 3	1, 0	1, 0	5, 0
5	1, 0	1, 0	3, 3	3, 0	1, 0	4, 2
6	4, 0	3, 3	4, 3	4, 0	1, 1	4, 2
7	1, 0	3, 0	5, 4	3, 0	2, 0	3, 1
8	-, -	-, -	-, -	-, -	-, -	-, -

arcs $[x, y]$ and all times t, let $l[x, y, t] = 0$ and $b[x, y, t] = -b(y, x, u)$, where $u = t + b(x, y, t)$. For all non-artificial arcs (x, y), we list their values of $b(x, y, t)$ and $l(x, y, t)$ in Table 3.1 (Assume $T = 8$).

Figure 3.12. The initial network of Example 3.4

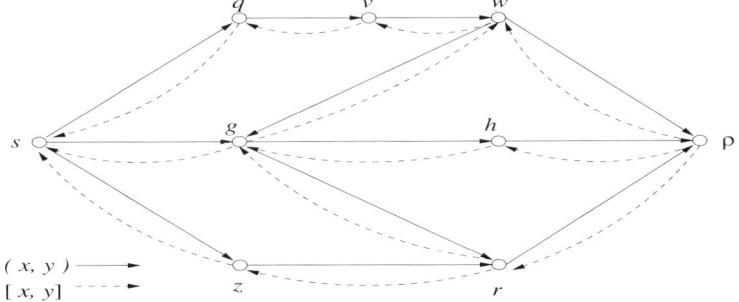

First, we set $d(s, t) = 0$ ($t = 0, 1, ..., T$) and all other labels $d(y, t) =$ *null.* $Q = \{d(s, 0), d(s, 1), ..., d(s, T)\}$. Consider $d(s, 0)$ first.

Since $b(s, q, 0) = 1$, $l(s, q, 0) = 1 > 0$, q can be labelled by setting $d(q, 1) = \{s\}$. Similarly, $d(g, 1)$ and $d(z, 1)$ can also be set to $\{s\}$. Delete $d(s, 0)$ from Q and append $d(q, 1)$, $d(g, 1)$ and $d(z, 1)$ in Q.

Consider $d(s, 1)$ (now it becomes the first element in Q). Noting $b(s, q, 1) = 1$, but $l(s, q, 1) = 0$, we cannot send any flow starting from time 1 through arc (s, q) since during this time the capacity of (s, q) is zero. Then $d(q, 2)$ remains null. For arc (s, g), since $b(s, g, 1) = 3$ and $l(s, g, 1) = 0$, g cannot be labelled at time 4 and $d(g, 4)$ remains null. Note that $d(g, 2)$ and $d(g, 3)$ remain null too, as no flow can arrive at g at time 2 or 3.

Following this process, we can label other vertices. When Q becomes empty, this iteration is completed. The result is shown in Figure 3.13 (all other labels $d(y, t)$ which do not appear in the figure are null).

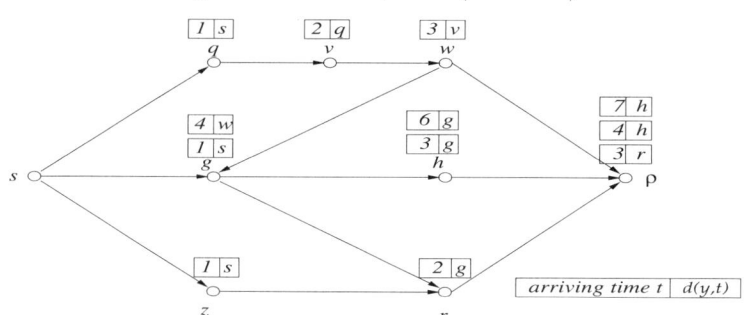

Figure 3.13. Example 3.4 (continued)

Note that $d(\rho, 3) = \{r\}$, $d(\rho, 4) = \{h\}$ and $d(\rho, 7) = \{h\}$. The earliest arrival time for vertex ρ is 3. By a backward searching, we can find a feasible dynamic f-augmenting path $P_1 = \{s, g, r, \rho\}$ with $\alpha(P_1) = 3$. Let f_1 denote the dynamic flow along P_1; then we have

$$v(f_1) = \min\{l(s, g, 0), l(g, r, 1), l(r, \rho, 2)\} = \min\{3, 2, 2\} = 2$$

$$\max f = v(f_1) = 2$$

And then, update N. Let

$$l(s, g, 0) := l(s, g, 0) - v(f_1) = 3 - 2 = 1$$

$$l[g, s, 1] := l[g, s, 1] + v(f_1) = 0 + 2 = 2$$

$$l(g, r, 1) := l(g, r, 1) - v(f_1) = 2 - 2 = 0$$

$$l[r, g, 2] := l[r, g, 2] + v(f_1) = 0 + 2 = 2$$
$$l(r, \rho, 2) := l(r, \rho, 2) - v(f_1) = 2 - 2 = 0$$
$$l[\rho, r, 3] := l[\rho, r, 3] + v(f_1) = 0 + 2 = 2$$

Other b and l remain unchanged.

We illustrate the labelling results in the remaining iterations in Figures 3.14-3.16. To highlight the previous feasible dynamic f-augmenting paths, we use dotted lines to represent them in each figure. The network updating process for each iteration is omitted.

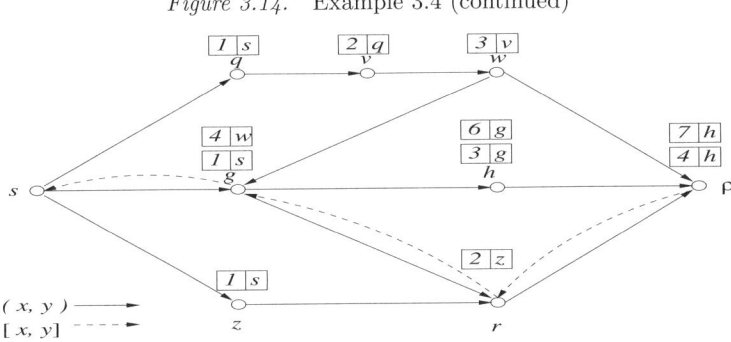

Figure 3.14. Example 3.4 (continued)

$$P_2 = \{s, g, h, \rho\} \qquad \alpha(P_2) = 4$$
$$v(f_2) = \min\{l(s, g, 0), l(g, h, 1), l(h, \rho, 3)\} = \min\{1, 2, 4\} = 1$$
$$maxf := maxf + v(f_2) = 2 + 1 = 3$$

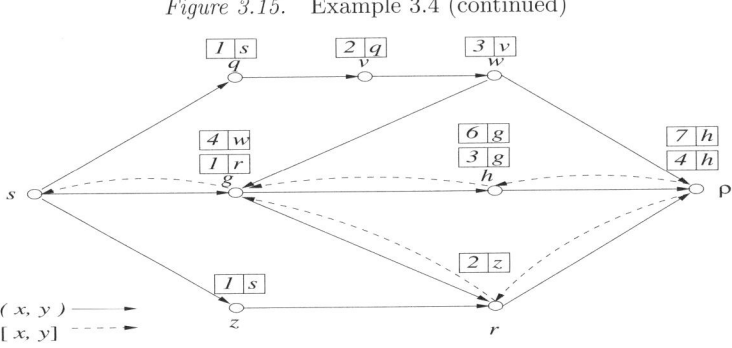

Figure 3.15. Example 3.4 (continued)

$$P_3 = \{s, z, r, g, h, \rho\} \qquad \alpha(P_3) = 4$$

$$v(f_3) = \min\{l(s, z, 0), l(z, r, 1), l[r, g, 2], l(g, h, 1), l(h, \rho, 3)\}$$
$$= \min\{2, 3, 2, 1, 3\} = 1$$
$$maxf := maxf + v(f_3) = 3 + 1 = 4$$

Figure 3.16.   Example 3.4 (continued)

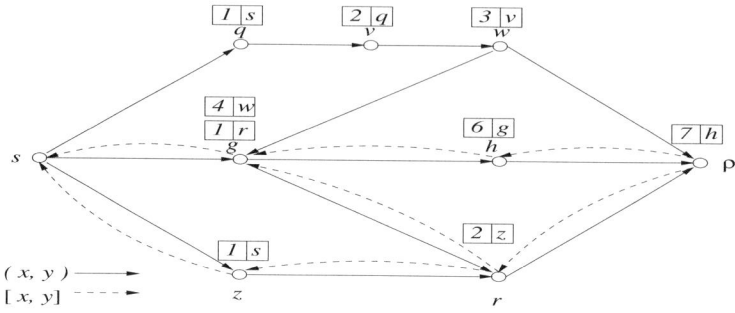

$$P_4 = \{s, q, v, w, g, h, \rho\} \qquad \alpha(P_4) = 7$$

$$v(f_4) = \min\{l(s, q, 0), l(q, v, 1), l(v, w, 2), l(w, g, 3), l(g, h, 4), l(h, \rho, 6)\}$$
$$= \min\{1, 2, 2, 1, 2, 1\} = 1$$
$$maxf := maxf + v(f_4) = 4 + 1 = 5$$

Figure 3.17.   Example 3.4 (continued)

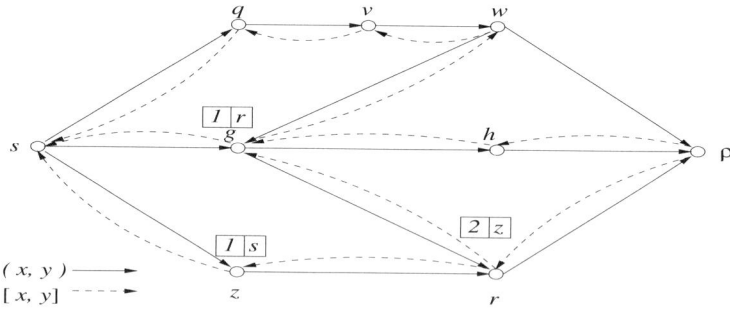

On the last iteration (see Figure 3.17), all labels $d(\rho, t) = null$ ($t = 0, 1, ..., T$); so $t_{min} = T+1$. Thus, the algorithm stops, and the universal maximum dynamic flow is $f = \{f_1, f_2, f_3, f_4\}$ with $v(f) = 5$.

Now let us examine the optimality and the time complexity of Algorithm TVUMF-ZW.

Theorem 3.5 *Algorithm TVUMF-ZW can optimally solve the time-varying universal maximum flow problem with no waiting time being permitted at any vertex.*

Proof. Note that in each iteration (i.e., an execution of the whole loop), Algorithm TVUMF-ZW either finds a feasible dynamic f-augmenting path $P(s, \rho)$ or stops as the sink vertex cannot be labelled. Assume that the algorithm stops at the $(j + 1)$th iteration (note that Algorithm TVUMF-ZW must stop in finite iterations since all arc capacities are nonnegative integers and each feasible dynamic f-augmenting path obtained by Algorithm TVUMF augments at least one unit of flow), and let λ be the schedule found by the algorithm to send flows over the period $[0, T]$. The total flow value under λ, $f(\lambda, T)$, is equal to $f(\lambda, T) = \sum_{t=0}^{T} g(\lambda, t)$, where $g(\lambda, t)$ denotes the value of all flows arriving at ρ at time t. Given a time t such that $0 \le t \le T$, let $f(\lambda, t) = \sum_{t' \le t} g(\lambda, t')$. Now, we prove that λ is the optimal schedule within the time t for all $0 \le t \le T$. In other words, if $F(t)$ is the maximum flow of N within time t, we need to prove that $F(t) = f(\lambda, t)$. Clearly, we have $F(t) \ge f(\lambda, t)$. Thus, it will suffice if we can show that, for any t, $F(t) \le f(\lambda, t)$.

We will prove the argument by induction on t ($0 \le t \le T$). Consider first $t = 0$. Since all transit times b are positive integers, no flow can be sent from s to ρ within time zero. Thus $F(t) = 0$ and $F(0) \le f(\lambda, 0)$ holds.

Assume that $F(t') \le f(\lambda, t')$ for all $t' < t$. We now prove $F(t) \le f(\lambda, t)$ under this assumption.

Let $\Delta F = F(t) - F(t - 1)$ and $\Delta f = f(\lambda, t) - f(\lambda, t - 1)$. Note that according to the definition of $f(\lambda, t)$, we have $\Delta f = g(\lambda, t)$ (recall that $g(\lambda, t)$ is the value of all flows arriving at ρ at time t). Then, by the induction assumption, we only need to prove $\Delta F \le \Delta f$. Suppose $\Delta F > \Delta f$. That is, $\Delta F = \sum_{x \in \aleph} F(x, \rho, \tau(x)) > \sum_{x \in \aleph} f(x, \rho, \tau(x)) = \Delta f$, where $\aleph = \{x | (x, \rho) \in A, \tau(x) + b(x, \rho, \tau(x)) = t\}$, $F(x, y, \tau)$ and $f(x, y, \tau)$ denote the flow values that arrive at y via the arc (x, y) departing at time τ, according to $F(t)$ and $f(\lambda, t)$, respectively. Then, there must exist at least one arc $(x_r, \rho) \in A$ such that $F(x_r, \rho, \tau(x_r)) > f(x_r, \rho, \tau(x_r))$. As any flow on N must satisfy the flow conservation condition at each intermediate vertex of N, there must exist at least one arc $(x_{r-1}, x_r) \in A$ such that $F(x_{r-1}, x_r, \tau(x_{r-1})) > f(x_{r-1}, x_r, \tau(x_{r-1}))$. Clearly, following this process, we can find a dynamic path $P = (s, x_1, x_2, ..., x_r, \rho)$, where $F(x, y, \tau(x)) > f(x, y, \tau(x))$ on each arc (x, y). Consequently, letting

$$\delta = \min_{\{(x,y) \text{ on } P\}} \{F(x, y, \tau(x)) - f(x, y, \tau(x))\},$$

we know there exists a path on the dynamic residual network, along which we can augment δ units of flow from s to ρ, where $\delta > 0$. This implies that $g(\lambda, t)$ is not the value of all possible flows that can arrive at ρ at time t, because it is still possible to augment an additional flow δ from s which arrive at ρ at time t. This is a contradiction. Therefore we must have $\Delta F \leq \Delta f$. This, together with the inductive hypothesis, gives us $F(t) \leq f(\lambda, t)$.   □

Theorem 3.6 *Algorithm TVUMF-ZW can be implemented in $O(Unm T^2)$ time.*

Proof. Consider one iteration. In the initialization block, the running time is $O(T)$. For the labelling operation, we may need to examine all arcs at all times t (in the worst case); so the running time is $O(Tm)$. The algorithm applies the procedure UPNET to update capacities, which needs a running time $O(Tm)$. Hence the total running time in one iteration is bounded by $O(Tm)$. Suppose U is the maximum capacity of arcs, then the maximum flow value of the dynamic network is bounded by $O(nTU)$, since at each time t ($0 \leq t \leq T$), there may be no more than n paths sending flows to ρ and the maximum flow on any possible path is at most U. Each iteration at least augments one unit of flow; so the algorithm will terminate within nTU iterations. Thus, the total running time of Algorithm TVUMF-ZW is bounded by $O(UnmT^2)$.   □

We can make some slight change to Algortihm TVUMF-ZW so that it can handle the case with $u_x = \infty$, i.e., the case where waiting at any vertex is arbitrarily allowed. Since a feasible f-augmenting path from s to y of time exactly t will be a feasible f-augmenting path from s to y of time exactly $t + 1$, if $l(y, t) > 0$, we can check whether there is still capacity available at y, namely, whether $l(y, t) > 0$, when we are to label y with $d(y, t + 1)$ after it is labelled with $d(y, t)$. In fact, if y is labelled with $d(y, t)$, then, for any $t' > t$, y can be labelled with $d(y, t')$ if $\prod_{\tau=t}^{t'-1} l(y, \tau) > 0$. On the other hand, if y is labelled with $d(y, t)$, then, for any $t'' < t$, y can be labelled with $d(y, t'')$ if $\prod_{\tau=t}^{t''+1} l[y, \tau] > 0$.

In view of the above, we can modify Algorithm TVUMF-ZW as follows to solve the problem with $u_x = \infty$ for each vertex $x \in V$.

Algorithm TVUMF-AW

All steps are same as those of Algorithm TVUMF-ZW, except the following labelling operation:

labelling:

If $(0 \leq t' \leq T)And(d(y,t') = null)And(c' > 0)$ **then**
    Let $\tau := t'; d(y,\tau) := x; Q := Q \cup \{d(y,\tau)\};$
    Let $\tau := \tau + 1;$
    While $(\tau \leq T)And(d(y,\tau) = null)And(l(y,\tau - 1) > 0)$ **do**
        Let $d(y,\tau) := x; Q := Q \cup \{d(y,\tau)\}; \tau := \tau + 1;$
    Let $\tau := t' - 1;$
    While $(\tau \geq 0)And(d(y,\tau) = null)And(l[y,\tau + 1] > 0)$ **do**
        Let $d(y,\tau) := x; Q := Q \cup \{d(y,\tau)\}; \tau := \tau - 1;$

Results similar to those for Algorithm TVUMF-ZW can be obtained on the optimality and time complexity of Algorithm TVUMF-AW.

Theorem 3.7 *Algorithm TVUMF-AW can optimally solve, in $O(Unm\,T^2)$ time, the TVUMF problem with no constraint on the waiting time at any vertex.*

The proof of Theorem 3.7 is similar to that of Theorem 3.5 and 3.6.

We now consider the case where a flow can wait at a vertex $x \in V$ subject to an upper bound u_x. First, we deal with the situation with no vertex capacity limit. Recall that Theorem 3.4 gives us a feasible condition for constructing a feasible f-augmenting path. To implement Theorem 3.4, consider the following scenarios:

(i) The vertex y is labelled with $d(y,t)$ through an arc (x,y) and t is the arrival time. Note that the waiting time at y is bounded by u_y, thus for $\tau = t + 1, ..., t + \alpha_{i^0}(x) + u_y$, all $d(y,\tau)$ can be labelled if $d(y,\tau) = null$ (note that here $l(y,\tau) = +\infty$). On the other hand, for any τ' such that $0 \leq \tau' \leq t$, y can be labelled with $d(y,\tau')$ when the condition $\prod_{\tau=t}^{\tau'+1} l[y,\tau] > 0$ is satisfied.

(ii) The vertex y is labelled with $d(y,t)$ through an artificial arc $[x,y]$, with $\tau(x)$ being the departure time at x and t being the arrival time. Since this is an artificial arc, $r_y^1[x,y,\tau(x)] \neq 0$, which means that there exists a previous augmenting flow that traverses the arc (y,x) with the departure time t. However, this flow may arrive at y before t and has a positive waiting time $r_y^1[x,y,\tau(x)]$ at y. Thus, by Theorem 4.4, for any $\tau = t + 1, ..., u_y + t - r_y^1[x,y,\tau(x)]$, y can be labelled with $d(y,\tau)$. On the other hand, for any τ' such that $0 \leq \tau' \leq t$, y can be labelled with $d(y,\tau')$ when $\prod_{\tau=t}^{\tau'+1} l[y,\tau] > 0$.

(iii) Suppose $d(y,t)$ has been labelled to y. Then, any non-artificial arc $(y,k) \in A$ can be examined similar to that in Algorithm TVUMF-ZW. For any artificial arc $[y,k] \in A$, only those $d(k,\tau)$ that satisfy the following conditions can be used to label k:

(1) $0 \leq \tau = t - r_y^1[y,k,u] + b[y,k,u] \leq T$ for a certain value of $b[y,k,u]$, where u satisfies $t = u + r_y^1[y,k,u];$

(2) $d(k, \tau) = null$;

(3) $l[y, k, u] > 0$.

We can sort u for $t = 0, 1, ..., T$ after applying the Network Update Procedure.

The following is the algorithm to solve the problem.

Algorithm TVUMF-BW

Begin

$maxf := 0$; $t_{min} := 0$;

While $t_{min} \leq T$ **do**

Initialization: $d(s, t) := 0, t = 0, 1, ..., T$; $Q := \{d(s, 0), d(s, 1), ..., d(s, T)\}$;

    While $Q \neq \emptyset$ **do**

      Select the first label in Q, denoted by $d(x, t)$;

      <u>checking:</u>

        For all y such that $(x, y) \in A$; or all y and u such that $[x, y] \in A, t = u + r^1_x[x, y, u]$ **do**

          Let $t' := t + b(x, y, t)$ (or $t' := t - r^1_x[x, y, u] + b[y, k, u]$ for each $b[y, k, u]$);

          If $t < t' \leq T$ **then** Let $c' := l(x, y, t)$ and $r^m_1 := 0$;

          Else Let $c' := l[x, y, u]$, $r^m_1 := r^1_y[x, y, u]$ and mark $l[x, y, u]$;

          <u>labelling:</u>

          If $(0 \leq t' \leq T)\text{And}(d(y, t') = null)\text{And}(c' > 0)$ **then**

            Let $\tau := t' - 1$;

            While $(\tau > 0)\text{And}(d(y, \tau) = null)\text{And}(l[y, \tau + 1] > 0)$ **do**

              Let $d(y, \tau) := x$; $Q := Q \cup \{d(y, \tau)\}$; $\tau := \tau - 1$;

            Let $\tau := t'$; $\zeta_2 := t'$;

            Do

              Let $\zeta_1 := \zeta_2$; $\zeta_2 := \max_{t_1 = \zeta_1 + 1, ..., \zeta_1 + u_y, t_1 \leq T} \alpha_1(y, t_1)$;

            While $(\zeta_2 > \zeta_1)$ And $(\zeta_2 + u_y \leq T)$;

            While $(\tau < \zeta_1 + u_y - r^m_1)\text{And}(\tau \leq T)$ **do**

              If $d(y, \tau) = null$ **then** $d(y, \tau) := x$; $Q := Q \cup \{d(y, \tau)\}$;

              Let $\tau := \tau + 1$;

         Let $Q := Q \backslash \{d(x, t)\}$;

    Let t_{min} be the minimum t such that $d(\rho, t) \neq null$. If $d(\rho, t) = null$ for all t, then let $t_{min} = T + 1$;

    If $t_{min} < T + 1$ **then**

      Use the predecessor indices to identify the feasible dynamic f-augmenting path $P(s, \rho)$ of time exactly t_{min};

      Let $g(t_{min}) :=$ the minimum capacity of arcs in $P(s, \rho)$;

      Augment $g(t_{min})$ units of flow along $P(s, \rho)$;

      Call revised procedure UPNET;

Table 3.2. u_x

vertex x	q	v	w	g	h	z	r
u_x	1	3	2	2	3	3	0

> Sort all values $u + r_x^1[x, y, u]$ for all $u = 1, 2, ..., T$ and all arc $[x, y]$;
> > **Let** $maxf := maxf + g(t_{min})$;
> **End.**

The following example is an illustration of the algorithm.

Example 3.5

The initial network in this example is same as that in Example 3.4, where the bounds of waiting times at the internal vertices are listed in Table 3.2.

In the first three iterations we find three feasible dynamic f-augmenting paths which are same as those in Example 3.4. Some labels $d(y, t)$, however, are different. The results are shown in Figures 3.18, 3.19 and 3.20 respectively.

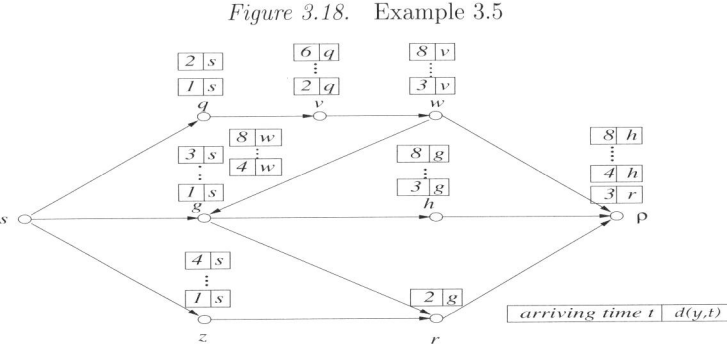

Figure 3.18. Example 3.5

Figure 3.21 shows the result of labels after the 4th iteration. Let us take a look of these labels $d(g, t)$. $d(g, 1)$ comes from $d(r, 2)$ since $b[r, g, 2] = -1$ and $l[r, g, 2] > 0$. Because $u_g = 2$, $d(g, 2)$ and $d(g, 3)$ can also be set to $\{r\}$. However, during the time period [1,3], no flow can be sent along arc (g, h). $d(g, 4)$ comes from $d(w, 3)$. Since $b(g, h, 4) = 2$ and $l(g, h, 4) > 0$, $d(h, 6)$ can be set to $\{g\}$. At the end of this iteration, we

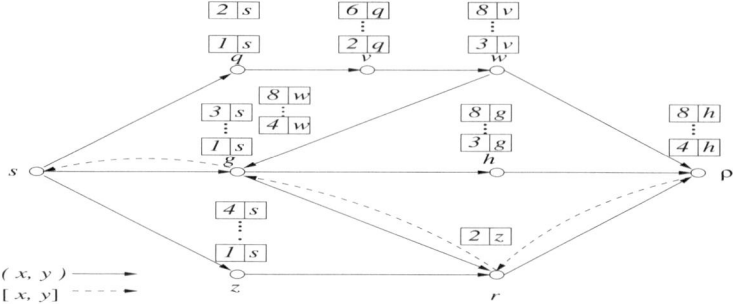

Figure 3.19.   Example 3.5 (continued)

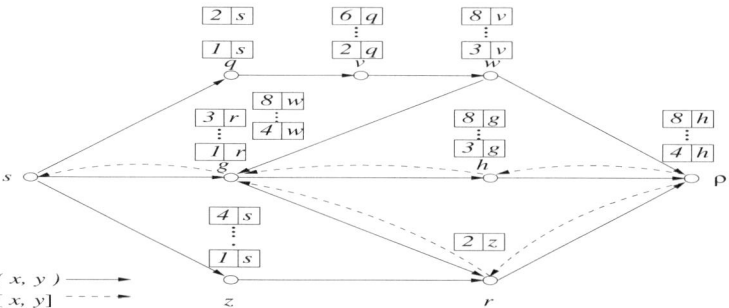

Figure 3.20.   Example 3.5 (continued)

find a feasible dynamic f-augmenting path $P_4 = \{s, q, v, w, g, h, \rho\}$ with $v(f_4) = 1$ (compare it with Example 3.4).

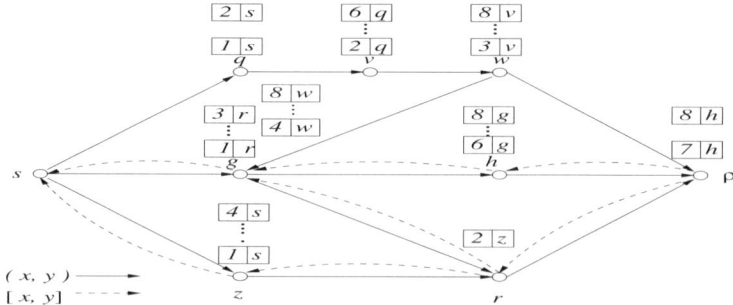

Figure 3.21.   Example 3.5 (continued)

The algorithm stops in the 5th iteration. Figure 3.22 shows all labels. The maximum flow is $f = \{f_1, f_2, f_3, f_4\}$ with $v(f) = 4$.

Figure 3.22. Example 3.5 (continued)

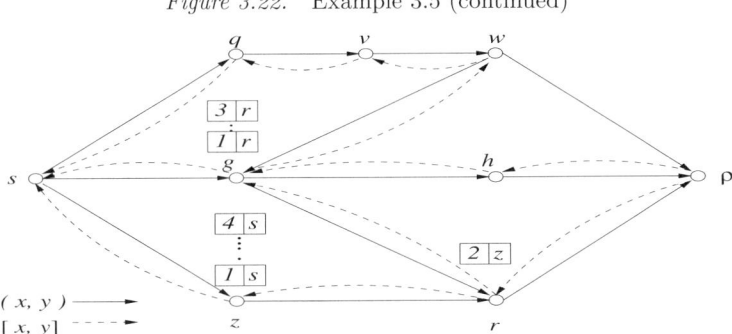

We now examine the optimality and time complexity of Algorithm TVUMF-BW.

Theorem 3.8 *Algorithm TVUMF-BW can optimally solve the TVUMF problem where the waiting time at each vertex is subject to an upper bound while the capacity at any vertex is unlimited.*

Proof. We use the same notions as those in the proof of Theorem 3.5. First, by Theorem 3.4, it is clear that the dynamic f-augmenting path obtained in each iteration of Algorithm TVUMF-BW is feasible under the bounded waiting time constraint. Let λ be the solution to send flows obtained by the algorithm. Then, λ is feasible. On the other hand, any feasible dynamic f-augmenting path can be found by the algorithm since Theorem 3.4 gives a sufficient and necessary condition. Next, noting that the conservation condition at each intermediate vertex and at each time t still holds, we can use a similar method as in the proof of Theorem 3.4 to find a feasible dynamic f-augmenting path of time t in the dynamic residual network created based on $f(\lambda, t)$. This will result in a contradiction to the fact that there is no dynamic f-augmenting path in the residual network. It indicates that $f(\lambda, t)$ must be a universal maximum flow in N. This completes the proof. $\qquad\square$

Theorem 3.9 *Algorithm TVUMF-BW can be implemented in $O(UnT^2 (m + n\log T))$ time.*

Proof. Consider one iteration. The initialization block needs a running time $O(T)$. For the labelling operation, we may need to examine all arcs at all times t (in the worst case), hence the running time is $O(Tm)$. For the revised procedure UPNET, we need a running time $O(Tn)$ for updating capacities and $O(\log u_x) = O(\log T)$ for inserting r_1 in R_1 to keep it in nondecreasing order. Thus, this step needs a running time $O(nT \log T)$, and therefore the total running time of one iteration is bounded by $O(T(m + n \log T))$. For the sorting we can use bucketsort, with T buckets. Since a dynamic path can contain at most $(n-1)T$ arcs, there exist at most $(n-1)T$ waiting times. So this step can be implemented in $O(nT)$. Suppose U is the maximum capacity of arcs. Then, the maximum dynamic flow value is bounded by $O(nTU)$. Since each iteration at least augments one unit of flow, the algorithm will terminate within $O(nTU)$ iterations. Thus, the total running time of Algorithm TVUMF-BW is bounded by $O(UnT^2(m + n \log T))$.   □

Now, we consider the case with vertex capacity constraints. Similar to the discussion for the problem under arbitrary waiting time constraints, we need to check whether $l(y, t) > 0$ when we are to label y with $d(y, t+1)$ after it is labelled with $d(y, t)$. In view of this, we modify Algorithm TVUMF-BW as follows:

Algorithm TVUMF-BW'
All steps are the same as those of Algorithm TVUMF-BW except the
following labelling operation:
labelling:
If $(0 \le t' \le T)$And$(d(y, t') = null)$And$(l' > 0)$ **then**
   Let $\tau := t' - 1$;
   While $(\tau > 0)$And$(d(y, \tau) = null)$And$(l[y, \tau + 1] > 0)$ **do**
      Let $d(y, \tau) := x$;
      Let $Q := Q \cup \{d(y, \tau)\}$;
      Let $\tau := \tau - 1$;
   Let $\tau := t'$;
   Let $\zeta_2 := t'$;
   Do
      Let $\zeta_1 := \zeta_2$;
      Let $\zeta_2 := \max_{t_1 = \zeta_1 + 1, \ldots, \zeta_1 + u_y, t_1 \le T} \alpha_1(y, t_1)$;
   While $(\zeta_2 > \zeta_1)$ And $(\zeta_2 + u_y \le T)$;
   If $(\tau < \zeta_1 + u_y - r_1^m)And(\tau \le T)And(d(y, \tau) = null)$ **do**
      Let $d(y, \tau) := x$;
      Let $Q := Q \cup \{d(y, \tau)\}$;
      Let $\tau := \tau + 1$;
   While $(\tau < \zeta_1 + u_y - r_1^m)And(\tau \le T)And(l(y, \tau - 1) > 0)$ **do**
      If $d(y, \tau) = null$ **then**
         Let $d(y, \tau) := x$;
         Let $Q := Q \cup \{d(y, \tau)\}$;
      Let $\tau := \tau + 1$;
End;

We can obtain results similar to those for Algorithm TVUMF-BW on the optimality and time complexity of Algorithm TVUMF-BW'.

We now turn to the few variants of the problem as listed at the end of Section 2, in this chapter. As we have mentioned, the optimal solution to send the maximum flow may not be unique. One may therefore like to find such a maximum flow solution that has the earliest (or the latest) arrival time at ρ or the earliest (or the latest) departure time at s, within the same time limit T. These cases are discussed below.

1. The earliest (or the latest) arrival maximum flow. To find the earliest arrival maximum flow, we can modify the algorithms described above, so that they always select the label $d(\rho, t)$ in increasing order of the time t. That is, from the feasible flows from s to ρ with different arrival times at vertex ρ, the flow with the earliest arrival time cam always be picked up first. Since any flow which arrives at ρ will never be changed by any f-augmenting paths found later, the maximum flow obtained with the above pickup mechanism will have the earliest arrival time at ρ.

Similarly, to find the latest arrival flow, we can modify the algorithms so that they select the label $d(\rho, t)$ in decreasing order of time t.

2. The earliest (or the latest) departure maximum flow. To find the flow that has the earliest departure time at s, we can modify the algorithm as follows: Let t_s denote the departure time at s; and attach it to the label $d(x, t)$ when x is labelled at time t. If $d(x, t)$ is labelled already, we retain the earliest (or the latest) one. By doing so, the flow finally generated will have the earliest (or the latest) departure time at s.

An interesting question is whether there exists a maximum flow that departs at s at the earliest possible time and arriving at ρ also at the earliest possible time. We have the following theorem.

Theorem 3.10 *For a given time-varying network N and a time duration T, there exist an optimal solution that transmits the maximum flow from the source vertex s at the earliest departure time and arrives at the sink vertex ρ at the earliest arrival time.*

Proof. Suppose that there are two maximum flows, F_1 and F_2, in N, within time T, with F_1 having its departure time earlier than that of F_2 but the arrival time later than that of F_2. Then, we can show that there must exist another maximum flow, denoted by F', which possesses the departure time of F_1 and the arrival time of F_2.

Decompose F_1 and F_2 into subflows such that each subflow traverses a dynamic path in N with unit flow value. Let $f_1^1, ..., f_{l_1}^1$ and $f_1^2, ..., f_{l_2}^2$ be those subflows. Therefore, there must have two subflows, say, f_i^1 and f_j^2, which traverse the same arc (x, y) at the time t_0 (Otherwise, if no two subflows share the same arc at the same time, we can combine F_1 and F_2 to obtain a larger flow. This contradicts to the fact that both of F_1 and F_2 are the maximum flow in N). Furthermore, Let t_s^1, t_s^2, t_ρ^1, and t_ρ^2 be the departure time and the arrival time of f_i^1 and f_j^2, respectively, with $t_s^1 < t_s^2$ and $t_\rho^1 > t_\rho^2$ (Otherwise, F_1 will have the arrival time earlier than that of F_2, which contradicts to the assumption). We further let $p_i(x, \rho)$ and $p_j(x, \rho)$ be the sections of the paths P_i and P_j that are traversed by f_i^1 and f_j^2. Then, we construct a new subflow f' by traverses $p_1(s, x)$ and $p_2(x, \rho)$. Thus, f' has the departure time t_s^1 and the arrival time t_ρ^2. Then, we construct two new paths, say P_i' and P_j', by exchanging two sections $p_i(x, \rho)$ and $p_j(x, \rho)$ on paths P_i and P_j. Still denote two new subflows which traverse on P_i' and P_j' as f_i^1 and f_j^2. Notice that now the arrive time at ρ of f_i^1 is earlier than that of f_j^2.

Repeating doing this till there are no subflows f_i^1 and f_j^2 with $t_s^1 < t_s^2$ and $t_\rho^1 > t_\rho^2$. Let F' be the flow obtained by uniting all subflows f_i^1. It can be seen that F' has the earliest departure time as that of F_1 and the earliest arrival time as that of F_2. This completes the proof.   □

Results similar to Theorem 3.10 can also be obtained for the other three variants of problems. We omit the details here.

7.   Additional references and comments

Ford and Fulkerson (1958,1962) introduce the concept of *dynamic flows* in a network and propose the *maximal dynamic network flow problem*. Although their model is not time-varying (all parameters are time-independent), it is widely regarded as the fundamental work on the time-varying maximum flow problem. Orlin (1983) considers the problem with an infinite time horizon and the flow is to be sent through the network in each period of time so as to satisfy the upper and lower bounds. He formulates the problem as an infinite integer program.

Bellmore and Vemuganti (1971) examine the multi-commodity maximum dynamic flow problem. Philpott (1990), Anderson and Philpott (1994) consider a continuous dynamic maximum flow model in which the arc capacities vary as Lebesgue-measurable functions of times, transit time of each arc is constant, and waiting at intermediate vertex is allowed. They generalize the max-flow min-cut theorem.

After Gale (1959) propose his model with time-varying arc capacities, Minieka (1974) discusses a special case of Gale's model, in which each arc capacity has two possible values, a normal value and a zero value (corresponding to, respectively, the situation where the arc is usable and unavailable). Halpern (1979) examines the problem in which the vertex capacities are also time-varying. Xue, Sun and Rose (1998) study the fast data transmission problem by formulating it as a dynamic maximum flow problem. Carey (1987) deals with a time-varying flow problem in which the time taken to traverse each arc depends on the flow rate on the arc.

Orda and Rom (1995) investigate another version of the time-varying maximum flow problem, in which all transit times, arc capacities and vertex capacities are time-dependent, and waiting at the intermediate vertices is allowed. They establish a generalized max-flow min-cut theorem for their model.

Anderson, Nash and Philpott (1982) study a continuous-time problem in which the transit time of each arc is constant, and both arcs and vertices are subject to capacity limits. The problem is formulated as an infinite linear program, and a continuous-time version of the the well-known labelling algorithm is proposed to solve this problem. Blum (1990, 1993) examines the issues of approximating a continuous max-flow problem by a sequence of static max-flow models in finite networks. He introduces a definition of a continuous flow and proves that the approximation sequence of network flows generated by his algorithm has a weakly convergent subsequence which converges to a maximal continuous flow. Jacobs and Seiffert (1983) address similar problems.

Chapter 4

TIME-VARYING MINIMUM COST FLOW PROBLEMS

1.   Introduction

The *minimum cost flow problem* has been extensively studied in the literature (see, e.g., Ahuja et al (1993)). The problem is to determine how a given amount of flow should be sent from one vertex (source) to another vertex (the sink) at minimum cost, subject to the capacity limits on the arcs of the network. Traditionally, this problem is considered as a static one, where it is assumed that it takes zero time to traverse any arc, and all attributes of the network, including the cost to send flow on an arc, and the capacity of an arc, are time invariant. In many practical situations, however, these assumptions are no longer valid. Clearly, a more realistic model is to take into account the time needed to traverse an arc.

We will address, in this chapter, the *time-varying minimum cost flow (TVMCF) problem*. The problem is to determine how to send a given amount of flow from the source vertex s to the sink vertex ρ before a pre-specified deadline T so as to minimize the total cost. This requires us to determine the best routes to send the flow from s to ρ and the best schedules to send the flow along these routes. Since the transit time, the cost, and the capacity on an arc are time varying, it may be necessary to wait at the starting vertex of the arc for the best departure time. Therefore, in addition to the routes to send the flow, waiting times at all vertices along each route will also be decision variables in our model.

The remainder of this chapter is organized as follows. In Section 2, we will introduce the basic formulation of the problem. A property on negative cycles will be described in Sections 3. Section 4 will be devoted to algorithm developments. Some special cases that can be solved more efficiently will be discussed in Section 5. As an application, we will study a time-varying maximum (f, c)-flow problem in Section 6, which can be solved by transformation to a time-varying minimum cost flow problem. Finally, some concluding remarks will be provided in Section 7.

2.   Concepts and problem formulation

Let $N(V, A, b, c, l)$ be a time-varying network, where all parameters are as defined in the previous chapters. Moreover, similar to Chapter 3, we define $f(x, y, t)$ as the flow travelling on the arc (x, y) during the period $[t, t + b(x, y, t)]$, $f(x, t)$ the flow waiting at vertex x during the period $[t, t + 1]$, and $f(\lambda, T)$ the total flow under a schedule λ, which specifies when and how to send flows from the source s to the sink ρ within the time limit T. Clearly,

$$f(\lambda, T) = \sum_{(x, \rho) \in A, t + b(x, \rho, t) \leq T} f(x, \rho, t).$$

The base problem addressed in this chapter is to find a feasible schedule λ to send a given flow v_f from s to ρ within the time limit T so as to minimize the total cost.

Without ambiguity, in the following we will assume that the length of an arc is equal to its cost, and use interchangeably the terminologies *cost* and *length*, and *shortest path* and *cheapest path*. Recall that, in Definition 3.2, we have defined the dynamic f-augmenting path, which is actually such a path that is feasible in terms of matching all the transit times and waiting times, and that can be used to transmit a positive flow as it has a positive capacity on its arcs and at its vertices.

The algorithms to be developed in this chapter will search, successively, shortest dynamic f-augmenting paths from the source vertex to the sink vertex in a dynamic residual network and then transmit as much as possible flow along the paths. Similar to Chapter 3, we will need a network updating procedure to retain the relevant information on the current dynamic flow. First, we create a new network to replace the original one:

For every arc $(x, y) \in A$, we create an artificial arc $[y, x]$. Its transit time $b[y, x, t]$, transit cost $c[y, x, t]$, and capacity $l[y, x, t]$ are defined as follows:

$$b[y, x, t] = \begin{cases} -b(x, y, u), & if \quad 0 \leq t = u + b(x, y, u) \leq T, u = 0, 1, ..., T \\ +\infty, & otherwise \end{cases}$$

$$c[y, x, t] = \begin{cases} -c(x, y, u), & if \quad 0 \leq t = u + b(x, y, u) \leq T, u = 0, 1, ..., T \\ +\infty, & otherwise \end{cases}$$

$$l[y, x, t] = 0; \quad \forall (x, y) \in A, t = 0, 1, ..., T.$$

For every vertex $x \in V$, we define $l[x, t]$ as the capacity within which a flow can be "stored" or "waiting" at x from time t to $t - 1$, and $c[x, t]$ as the cost for the flow to stay at x from time t to $t - 1$. Initially, let $l[x, t] = 0$ and $c[y, t] = -c(y, t - 1)$ for all x and t.

Originally, no flow can be sent along any artificial arcs in the network as defined above since capacities of those arcs are set to zero. Hence, this new network is equivalent to the original one, and so we will still denote it by N. After a feasible f-augmenting path is found, we create a dynamic residual network by using the procedure UPNET (see Section 3, Chapter 3). The optimization problems in the original network and in the dynamic residual network are equivalent in the sense that there is a one-to-one correspondence between their feasible solutions. Notice that, in the original network, we assume that all transit times $b > 0$. Thus, the first dynamic f-augmenting path will only contain arcs with positive transit times b and positive waiting times w. But in a dynamic residual network, the transit time associated with an artificial arc is a negative number, and a flow can be stored at a vertex for a negative waiting time. Therefore, a dynamic f-augmenting path found in the dynamic residual network may contain some arcs with negative transit times and negative waiting times. Accordingly, we need a definition to define the cost of a dynamic f-augmenting path while considering the cost of those artificial arcs.

Definition 4.1 *Let $P(x_1, ..., x_r)$ be a dynamic f-augmenting path from x_1 to x_r. Let*

$$W(x) = \begin{cases} \sum_{t'=0}^{w(x)-1} c(x, t' + \alpha(x)), & if \quad \alpha(x) < \tau(x) \\ \sum_{t'=0}^{|w(x)|-1} c[x, -t' + \alpha(x)], & if \quad \alpha(x) > \tau(x) \end{cases}$$

denote the waiting cost at x on P. Further, let $\zeta(x_1) = W(x_1)$, and define recursively

$$\zeta(x_i) = \begin{cases} \zeta(x_{i-1}) + c(x_{i-1}, x_i, \tau(x_{i-1})) + W(x_i), & if \quad (x_{i-1}, x_i) \in A^+ \\ \zeta(x_{i-1}) + c[x_{i-1}, x_i, \tau(x_{i-1})] + W(x_i), & if \quad [x_{i-1}, x_i] \in A^- \end{cases}$$

for $i = 2, ..., r$. The cost of P, $\zeta(P)$, is defined as $\zeta(x_r)$.

In Section 4, we will present our algorithms to solve the time-varying network problem we formulate above. We will examine three versions of the problem, which correspond, respectively, to the three situations regarding waiting at a vertex; namely, waiting is prohibited; arbitrarily allowed; and subject to an upper bound. Before we proceed, let us present another approach, which can also be used, theoretically, to solve the TVMCF problem, albeit with very high time complexity. This is given in Remark 4.1 below.

Remark 4.1 By utilizing the discrete-time feature of the time-varying network, the TVMCF problem can be re-formulated as a linear optimization model, and then solved by applying some standard optimization method, such as linear programming (LP) or dynamic programming (DP). To be more specific, let us consider our time-varying problem with arbitrary waiting times at vertices as an example. This problem can be written as:

$$\min \quad \sum_{(x,y)\in A}\sum_{t} c(x,y,t)f(x,y,t) + \sum_{x\in V}\sum_{t} c(x,t)f(x,t)$$

$$s.t. \quad \sum_{(s,x)\in A}\sum_{t} f(s,x,t) = v \tag{4.1}$$

$$\sum_{(x,y)\in A, t'+b(x,y,t')=t} f(x,y,t') + f(y,t) - \sum_{(y,x)\in A} f(y,x,t) = 0$$

$$\forall y \in V\backslash\{s,\rho\}, \quad t = 0,1,...,T \tag{4.2}$$

$$\sum_{(x,\rho)\in A}\sum_{0\leq t\leq T, t+b(x,\rho,t)\leq T} f(x,\rho,t) = v \tag{4.3}$$

$$0 \leq f(x,y,t) \leq l(x,y,t), \quad \forall(x,y)\in A, \quad t = 0,1,...,T$$

$$0 \leq f(x,t) \leq l(x,t), \quad \forall x\in V, \quad t = 0,1,...,T$$

The above can be solved by a standard LP algorithm. However, a difficulty with this approach is its excessive time requirement. For example, the time complexity of an efficient LP algorithm is $O(MN^{9/2})$ (see, e.g., Ye (1997)), where M is the number of constraints and N is the number of decision variables. This time complexity becomes to $O(T^{11/2}(m+n)^{11/2})$ when it is applied to the model above (as $M = N = (m+n)T$).

Remark 4.2 It is well-known that the static version of the minimum cost flow problem is polynomially solvable. The time-varying version of the problem is, however, NP-hard in ordinary sense. This can be seen from the fact that the time-varying shortest path problem is a special case of TVMCF, but it is an NP-hard problem (see Chapter 1). The TVMCF problem is NP-hard in the ordinary sense, since it is solvable in pseudo-polynomial time (see the next section).

3.   On the negative cycle

As we have mentioned above, we will solve the problem by searching for the shortest dynamic f-augmenting path in the residual network successively. To ensure that the algorithm can be finished within finite steps, we need a condition that the network should not contain any "negative cycle". The following definition introduce the concept of negative cycle in a time-varying network.

Definition 4.2 *A dynamic path $P(x_1, ..., x_r)$ is called a dynamic cycle if $x_1 = x_r$ and one can traverse this path starting from x_1 at a time t and returning to $x_r = x_1$ at the same time t. A negative cycle is defined as such a dynamic cycle whose total cost is negative and whose capacity is greater than zero.*

It is clear that, if a network, either the original network or the residual network, contains a negative cycle and it can be reached from s, then one can continuously travel along this cycle while the cost is decreased unlimited. In such a case, Thus, the problem has no optimal solution.

The original network contains no negative cycle, since all arcs in A^+ have positive transit times and all arcs in A^- have zero capacities. Now examine the residual network. Recall that a residual network is generated based on a flow sent from s to ρ. For a general flow, the generated residual network may have negative cycles. However, if the flow is sent along a shortest dynamic f-augmenting path, the residual network generated will contain no negative cycle. This can be seen as follows: Suppose that f is a flow sent along a shortest f-augmenting path P and C is a negative cycle in the residual network N' generated by f. Clearly, C and P must have common sections. Consider the case where they have one common section. Note that this common section will have opposite directions (refer to Sections 3, Chapter 3 and Section 2, Chapter 4). Denote S_c and S_p as the sections in cycle C and in path P, respectively, then we have $\zeta(S_c) = -\zeta(S_p)$. Let S' be the remaining section of C. Note that the cost of the cycle $\zeta(C) = \zeta(S') + \zeta(S_c) < 0$, i.e., $\zeta(S') < -\zeta(S_c) = \zeta(S_p)$. Therefore, replacing S_p by S' in P, we can generate a path that is shorter than P. This contradicts the assumption that P is a shortest dynamic f-augmenting path.

The analysis above is summarized in the following property.

Property 4.1 *If a time-varying network N contains no negative cycle, then the residual network generated based on a shortest dynamic f-augmenting path contains no negative cycle.*

Proof. Let N' be the residual network generated based on a flow f sent along a shortest dynamic f-augmenting path P in N. Suppose that N' contains negative cycles with $C = (x_0, y_1, ..., y_l, x_r, x_{r-1}, x_1, x_0)$ as the one with the minimum cost. Since C is generated based on f, C and P must have common sections.

First, we consider the case where P and C have one common section. The dotted line in Figure 4.1 represents the shortest dynamic f-augmenting path $P(s, \rho) = (s, ..., x_0, x_1, ..., x_r, ..., \rho)$, where $S_p = (x_0, x_1, ..., x_r)$ is a section of P. $S_c = (x_r, x_{r-1}, ..., x_0)$ is the section of C in N' (see the solid line in Figure 4.1). Since S_p and S_c have opposite directions, we have $\zeta(S_c) = -\zeta(S_p)$. Let S' be the section $(x_0, y_1, ..., y_l, x_r)$. Since $\zeta(C) = \zeta(S') + \zeta(S_c) < 0$, we have $-\zeta(S_c) > \zeta(S')$. On the other hand, since $\zeta(S_p) = -\zeta(S_c)$, we have $\zeta(S_p) > \zeta(S')$. Noting that $P(s, \rho)$ is the shortest f-augmenting path from s to ρ in N' and both S_p and S' exist in N, we can use S' to replace S_p in $P(s, \rho)$ to obtain another path $P'(s, \rho)$, with $\zeta(P') < \zeta(P)$. This is a contradiction to the assumption that P is the shortest dynamic f-augmenting path. Therefore, N' can not contain any negative cycles.

Figure 4.1.   A negative cycle (case I)   *Figure 4.2.*   A negative cycle (case II)

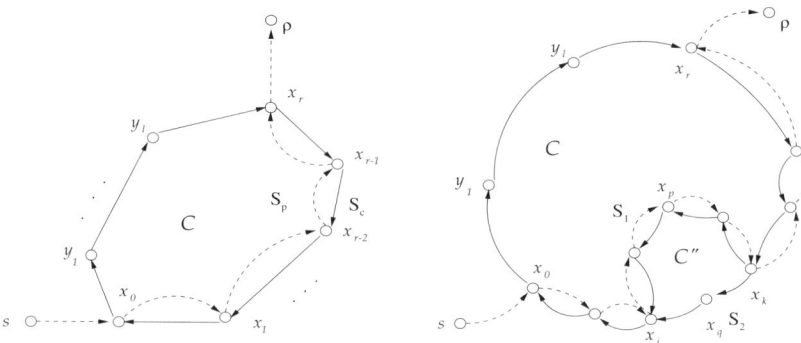

Now, we prove the claim that if $P(s, \rho)$ and C have more than one common section, then there must exist another cycle in N', say C', which has only one common section with P, and C and C' have the same cost. We will prove the case where P and C have two common sections (other cases can be dealt with in a similar way).

Suppose that P and C have two common sections $(x_0, ..., x_i)$ and $(x_k, ..., x_r)$. Then, there is another cycle $C'' = (x_i, ..., x_p, ..., x_k, ...x_i)$ in N' (see Figure 4.2). Let $S_1 = (x_i, ..., x_p, ..., x_k)$ and $S_2 = (x_k, ..., x_q, ..., x_i)$. Since both sections S_1 and S_2 exist in N, C'' exists in N too. Thus, C'' must have a non-negative cost, i.e., $\zeta(C'') = \zeta(S_1) + \zeta(S_2) \geq 0$. If $\zeta(C'') > 0$, we have $-\zeta(S_1) < \zeta(S_2)$. Then, replacing S_2 by the section $(x_k, ..., x_p, ..., x_i)$ in C will create a new cycle, with cost less than $\zeta(C)$. However, we have assumed that C is the minimal one among all negative cycles. Thus we must have $\zeta(C'') = 0$, i.e., $-\zeta(S_1) = \zeta(S_2)$. Then, we can use the section $(x_k, ..., x_p, ..., x_i)$ to replace section S_2 in C and form a new cycle C'. Note that C' and P have one common section only, a case we have proved in the above.

In summary, the proof is completed. □

4. Successive improvement algorithms

We will describe, in this section, algorithms that can find optimal solutions for the TVMCF problem. The algorithms are successive improvement procedures, which utilize the basic idea that, at each step, a shortest f-augmenting path is identified and then a certain amount of flow will be transmitted over the path. Since there is only a limited amount of flow to be transmitted from the source to the sink in the TVMCF problem, the algorithms will converge as long as we can show that at each step they can transmit a positive amount of flow over a shortest (cheapest) path.

We will apply the shortest path algorithms developed in Chapter 1 to identify the shortest f-augmenting paths. Corresponding to the three shortest path algorithms proposed in Chapter 1, we will consider, in Sections 4.1-4.3 below, three versions of the TVMCF problem, which deal respectively with the three types of waiting constraints at a vertex.

4.1 Waiting at any vertex is prohibited

Since no flow is allowed to wait at any vertex, we need not consider the waiting cost in this case.

As indicated above, the main idea to generate the solution is to find, repeatedly, the shortest dynamic f-augmenting path in the dynamic residual network. Note that in a dynamic residual network, the transit times may be positive or negative. To tackle this problem, we will develop a procedure, which contains two searching operations: *forward searching* and *backward searching*. Both operations are designed based on the idea of dynamic programming for the shortest path problem in Chapter 1. The *forward searching* is to deal with positive transit times, while the *backward searching* will deal with negative transit times.

The procedure to be developed will solve the following subproblem, where transit times and costs can be negative. Note that the concept of *nonnegative cycle* means a closed path with negative total travel time.

Subproblem SP1 – *Given a network N which contains nonzero transit times, arbitrary costs, and no negative cycles, find a shortest dynamic f-augmenting path from s to ρ with time at most T, where no waiting is permitted at any vertex.*

We also need the following definitions.

Definition 4.3 *Let $P(s = x_1, ..., x_r = x)$ be a dynamic f-augmenting path from s to x. A section $P(x_i, x_j)$, $1 \leq i < j \leq r$, is defined as a subpath of $P(s, x)$ provided that all the transit times on $P(x_i, x_j)$ have* *the same sign. A section of $P(s,x)$ is said to be positive (or negative) if its transit times are all positive (or negative).*

We can see that a dynamic f-augmenting path will consist of several positive and negative sections in an alternate manner. The number of these sections is said to be the *alternating number of $P(s,x)$*.

Definition 4.4 *Define $d_z(x,t)^k$ as the length of a shortest dynamic f-augmenting path from s to the vertex x of time exactly t with the alternating number at most k.*

Property 4.1 has indicated that neither the original network N nor any dynamic residual network contains any negative cycle. Consequently, we can show that a shortest dynamic f-augmenting path P contains no more than n vertices and each vertex cannot be visited more than once at any time t, $t = 0, ..., T$. Therefore, P cannot contain more than nT sections. In other words, $d_z(x,t)^k$ is the length of the shortest dynamic f-augmenting path from s to x of time exactly t when $k \geq nT$.

The procedure SDFP-ZW to solve the subproblem SP1 is presented below, where A^+ and A^- denote, respectively, the set of positive and the set of negative arcs.

Procedure SDFP-ZW
Begin
  Initialize: $d_z(s,0)^0 := 0, d_z(s,t)^0 := +\infty, t = 1, ..., T; d_z(y,t)^0 := +\infty, \forall y \in V \backslash \{s\}; l = 0, ..., T;$
  Sort all values $u + b(x,y,u)$ for $1 \leq u \leq T$ and for all arcs $(x,y) \in A^+$;
  Sort all values $u + b[x,y,u]$ for $0 \leq u \leq T - 1$ and for all arcs $[x,y] \in A^-$;
  $i := 0;$
  Do
    $i := i + 1;$
    For all $y \in V, t = 0, ..., T$ **do** $d_z(y,t)^i := d_z(y,t)^{i-1};$
    Case 1: i is an odd number:
      For $t = 1, ..., T$ **do**
        For every $y \in V \backslash \{s\}$ **do** *forward searching operation:*

$$d_z(y,t)^i := \min\{d_z(y,t)^i,$$

$$\min_{\{x|(x,y)\in A^+\}} \min_{\{u|u+b(x,y,u)=t \wedge l(x,y,u)>0\}} \{d_z(x,u)^i + c(x,y,u)\}\};$$

    Case 2: i is an even number:
      For $t = T - 1, ..., 0$ **do**

For every $y \in V \backslash \{\rho\}$ **do** *backward searching operation:*

$$d_z(y,t)^i := \min\{d_z(y,t)^i,$$

$$\min_{\{x|(x,y)\in A^-\}} \min_{\{u|u+b[x,y,u]=t \wedge l[x,y,u]>0\}} \{d_z(x,u)^i + c[x,y,u]\}\};$$

While there exists at least one $d_z(y,t)^i \neq d_z(y,t)^{i-1}$;
Let $d_z^*(\rho) := \min_{0 \le t \le T} d_z(\rho,t)^i$;
End

<u>Lemma 4.1</u> *When the procedure SDFP-ZW is terminated, $d_z^*(\rho)$ is the length of a shortest dynamic f-augmenting path from s to ρ with time at most T.*

Proof: We only need to prove that for each i and t, $d_z(y,t)^i$ obtained by the procedure is the length of a shortest dynamic f-augmenting path from s to y of time exactly t with the alternating number at most i. Note that any shortest path must contain a positive section as its first section since there are no negative arcs $(s,y) \in A^-$ in the original network N or any residual networks. Thus, we only need to consider paths whose first sections are positive.

The proof is carried out by double inductions on i and t. Consider $i = 1$. Use the second induction on time t. When $t = 0$, since $i = 1$ is an odd number, no positive dynamic path $P(s,y)$ of time exactly t exists in N except when $y = s$. We know that the length of the shortest dynamic path $P(s,s)$ of time 0 is 0. In the initialization of procedure SDFP-ZW, we have $d_z(y,0)^1 = +\infty$ ($y \in V \backslash s$) and $d_z(s,0)^1 = 0$; hence the claim holds.

Assume $t > 0$ and, for all values $t' < t$, $d_z(y,t')^1$ is the length of a shortest dynamic path $P(s,y)$ of time exactly t' with the alternating number at most 1 for all vertices y.

Consider a vertex y. First we prove that there exists a path of time exactly t with the alternating number 1 and with length $d_z(y,t)$. If $d_z(y,t)^1 = +\infty$, there is nothing to prove. So assume $d_z(y,t)^1$ is finite. Then by the forward searching operation, $d_z(y,t)^i$ comes from $d_z(x,u)^1 + c(x,y,u)$ for some x such that $(x,y) \in A^+$ and some u such that $u + b(x,y,u) = t$. By the induction on t, we know that there is a path $P'(s = x_1, ..., x_{r-1} = x)$ of time exactly u with the alternating number at most 1 and with length $d_z(x,u)^1$. Then, we extend the path with vertex y, obtaining a path $P(s,y)$. The time of $P(s,y)$ is exactly t and the length is $d_z(x,u)^1 + c(x,y,u) = d_z(y,t)^1$. This is what we want to obtain.

We now prove that $d_z(y,t)^1$ is the length of a shortest path from s to y of time exactly t. Let $P(s = x_1, ..., x_r = y)$ be a shortest path of time exactly t with the alternating number at most 1. If P is an empty path, then $y = s$ and the time of P is zero. The value $d_z(s,0)^1 = 0$ is correct.

Assume the path is not empty. Let x be the predecessor of y on this path. Let u be the time of the subpath $P(s,x)$, and let $\zeta(x)$ be the length of $P(s,x)$. By definition, $t = u + b(x,y,u)$. Since $u < t$, by induction on t, $\zeta(x) \geq d_z(x,u)^1$. By the calculation formula, P is a path of shortest possible length and of time exactly t, since there exists a path of time exactly t that achieves the length $d_z(y,t)^1$, as we showed above. This completes the proof for time t on $i = 1$.

Assume that for $i < k$, the claim is true. Now consider $i = k$. We discuss two cases in which i is an odd number and an even number, respectively.

Suppose i is an odd number. Consider $t = 0$. Since no negative cycles in N, the length of the shortest dynamic path $P(s,s)$ of time exactly 0 with the alternating number at most i is 0. For vertex $y \neq s$, since no flow can depart from y at time 0, no residual network can contain negative arcs which allow a flow to reach y from any other vertex x at time 0. This means there is no dynamic path $P(s,y)$ of time exactly 0. On the other hand, in the procedure, we let $d_z(y,0)^i = d_z(y,0)^{i-1}$ at first. By the induction on i we know $d_z(y,0)^{i-1} = +\infty$ and by the formula of the forward searching operation, $d_z(y,0)^i$ is unchanged. Thus $d_z(y,0)^i = +\infty$.

Assume the claim holds for $t' < t$. Now consider the case at the time t. One can see that there exists a path of time exactly t with the alternating number at most i and with the length $d_z(y,t)^i$. We now prove that $d_z(y,t)^i$ is the length of a shortest path from s to y of time exactly t. Let $P(s = x_1, ..., x_r = y)$ be a shortest path of time exactly t with the alternating number at most i. If P is an empty path, then $y = s$ and the time of P is zero. The value $d_z(s,0)^i = 0$ is correct.

Assume that P is not empty. Let x be the predecessor of y on this path. Let u be the time of the subpath $P(s,x)$, and let $\zeta(x)$ be the length of $P(s,x)$. By definition, $t = u + b(x,y,u)$. By induction, since $u < t$, $\zeta(x) \geq d_z(x,u)^i$. By definition, $P(s,y)$ is a path of shortest possible length and of time exactly t, and since there exists a path of time exactly t that achieves the length $d_z(y,t)^i$. This completes the proof.

The proof for the case with an even i can be conducted out in a similar way by using induction on k (let $t = T - k$, $k = 0, ..., T$). $\qquad\square$

Lemma 4.2 *The procedure SDFP-ZW can be implemented in $O(mnT^2)$ time.*

Proof: The time requirement of the initialization step is bounded by $O(nT)$. For the sorting we can use bucketsort, with T buckets. Since there are Tm values to be sorted, this step can be implemented in $O(Tm)$ time. One can see from the iterative formula that the number of iterations for i is proportional to $T \sum_x \sum_{y,(y,x)\in A} 1 = mT$. Since $i \leq nT$, the total time is bounded by $O(mnT^2)$. $\square$

Now we can describe the algorithm to solve the TVMCF problem with no waiting time permitted at any vertices. In the algorithm presented below, v represents the given flow value to be sent from the source to the sink, and f_j the maximal flow value which can be sent along the jth f-augmenting path $P_j(s,\rho)$.

> **Algorithm TVMCF-ZW**
> **Begin**
> $\bar{v} := 0$;
> **For** $j = 1, ..., v$ **do**
> **Call** procedure SDFP-ZW;
> **If** $d^*(\rho) < +\infty$ then call the procedure UPNET; (there is an f-augmenting path $P_j(s,\rho)$ with flow value $f_j = Cap(P_j(s,\rho))$; so update the network)
> **Else** stop; (no feasible solution to send all flow value v from s to ρ within time T)
> $\bar{v} := \bar{v} + f_j$;
> **If** $\bar{v} \geq v$ then stop;
> **End**

The algorithm is designed for seeking the optimal f-augmenting path from s to ρ. The real path can be obtained by a backtracking process.

To illustrate how the algorithm works, an example is given below.

Example 4.1

The original network is shown in Figure 4.3(a). Since no waiting time is permitted, all $l(x,t) = 0$. To simplify the figure, artificial arcs $[x,y]$ are not depicted there. Besides, in Figure 4.3(b) we only list those numbers corresponding to non-zero arc capacities. Suppose $v = 2$ and $T = 5$. The iterative process of Algorithm TVMCF-ZW for solving this problem is as follows:

$j = 1$; $i = 1$: all $d_z(y,t)^1$ are listed in Figure 4.4(b) (the blank spaces represent $+\infty$).

Figure 4.3.   Example 4.1

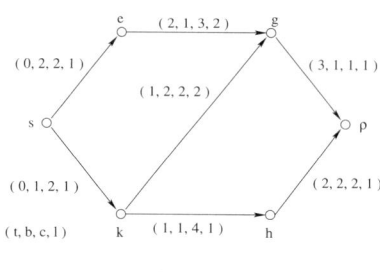

arc	t	b	c_t	1
(s, e)	0	2	2	1
(s, k)	0	1	2	1
(e, g)	2	1	3	2
(k, g)	1	2	2	2
(k, h)	1	1	4	1
(g, ρ)	3	1	1	1
(h, ρ)	2	2	2	1

(a)   (b)

Figure 4.4.   Example 4.1 (continued)

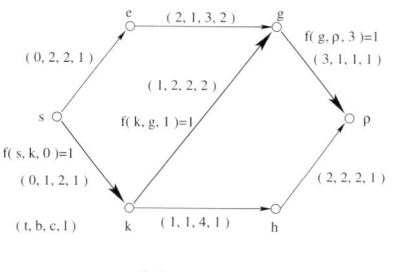

t	s	e	k	g	h	ρ
0	0					
1	0		2			
2	0	2			6	
3	0			4		
4	0					5
5	0					
min	0	2	2	4	6	5

(a)   (b)

$i = 2$: Since $A^- = \emptyset$, all $d_z(y,t)^2 = d_z(y,t)^1$, and the searching in this segment is stopped. We find a shortest dynamic f-augmenting path $P_1(s, \rho) = (s, k, g, \rho)$ with $f_1 = 1$. $\zeta(P_1) = d^*(\rho) = 5$ (see Figure 4.4(a)). We now execute the procedure UPNET and obtain a new network as shown in Figure 4.5(a).

Figure 4.5.   Example 4.1 (continued)

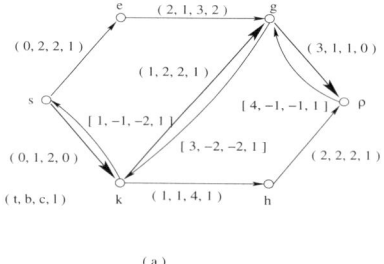

arc	t	b	c_t	1
(s, e)	0	2	2	0
[e, s]	1	−2	−2	1
(s, k)	0	1	2	1
(e, g)	2	1	3	2
(k, g)	1	2	2	1
[g, k]	3	−2	−2	2
(k, h)	1	1	4	1
(c, ρ)	3	1	1	0
[ρ, g]	4	−1	−1	1
(h, ρ)	2	2	2	1

(a)   (b)

$j = 2$; $i = 1$: all $d_z(y,t)^1$ are listed in Figure 4.6(b).

Figure 4.6. Example 4.1 (continued)

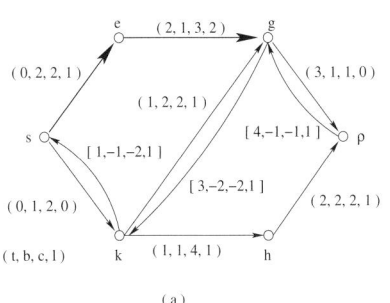

t	s	e	k	g	h	ρ
0	0					
1	0					
2	0	2				
3	0			5		
4	0					
5	0					
min	0	2		5		

(a)   (b)

$i = 2$: all $d_z(y,t)^2$ are listed in Figure 4.7(b) (notice that an artificial arc $[g,k]$ is added in path (s,e,g,k), see Figure 4.7(a)).

Figure 4.7. Example 4.1 (continued)

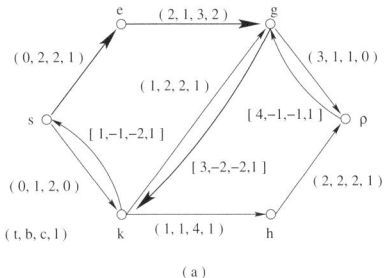

t	s	e	k	g	h	ρ
0	0					
1	0		3			
2	0	2				
3	0			5		
4	0					
5	0					
min	0	2	3	5		

(a)   (b)

$i = 3$: all $d_z(y,t)^3$ are listed in Figure 4.8(b).

$i = 4$: Since all $d_z(y,t)^4 = d_z(y,t)^3$, the searching in this segment is stopped. We find a shortest dynamic f-augmenting path $P_2(s,\rho) = (s,e,g,k,h,\rho)$ with $f_2 = 1$ and $\zeta(P_2) = d^*(\rho) = 9$ (see Figure 4.8(a)). Because $f_1 + f_2 = 2 = v$, the algorithm stops. The total cost is $\zeta(P_1) + \zeta(P_2) = 5 + 9 = 14$.

Theorem 4.1 *Algorithm TVMCF-ZW solves optimally the time-varying minimum cost flow problem with no waiting times at vertices.*

Proof: Straightforward.   □

From Lemma 4.2 and Algorithm TVMCF-ZW, we can easily obtain the following result.

Theorem 4.2 *The running time of Algorithm TVMCF-ZW is bounded above by $O(vmnT^2)$.*   □

Figure 4.8. Example 4.1 (continued)

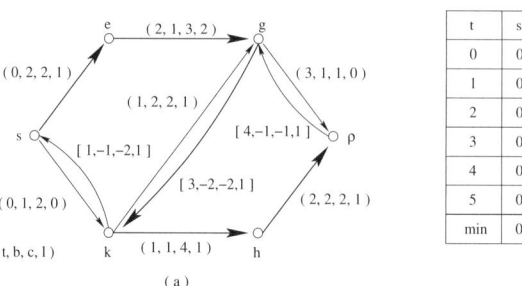

t	s	e	k	g	h	ρ
0	0					
1	0		3			
2	0	2			7	
3	0		5			
4	0					9
5	0					
min	0	2	3	5	7	9

(a)   (b)

4.2   Waiting at any vertex is arbitrarily allowed

We now consider the problem where waiting at any vertex is not subject to any constraints. Similar to Section 4.1, we will present a procedure, SDFP-AW, to solve a subproblem as follows:

Subproblem SP2 – *Given a network N with nonzero transit times, arbitrary costs, and no negative cycles, find a shortest dynamic f-augmenting path from s to ρ within time T, where waiting at any vertex is not limited.*

Note that, unlike the case with waiting time prohibited, a dynamic f-augmenting path $P(s, x)$ of time at most t will be a path of time at most $t + 1$ if $l(x, t) > 0$, since the flow can wait at x from t to $t + 1$. On the other hand, if $\alpha(x) = t$ in $P(s, x)$ and $l[x, t] > 0$, then a flow can have a negative waiting time $w(x) = -1$ at vertex x and the arrival time x can be $t - 1$.

<u>Definition 4.5</u> *Define $d_a(x, t)^k$ as the length of a shortest dynamic f-augmenting path from s to vertex x of time at most t, with the alternating number at most k.*

The following is our procedure to solve the subproblem SP2:

> **Procedure SDFP-AW**
> **Begin**
> **Initialize:** $d_a(s, t)^0 := 0, d_a(y, t)^0 := +\infty, \forall y \in V \backslash \{s\}; t = 0, ..., T;$
>   **Sort** all values $u + b(x, y, u)$ for $1 \leq u \leq T$ and for all arcs $(x, y) \in A^+;$
>   **Sort** all values $u + b[x, y, u]$ for $0 \leq u \leq T - 1$ and for all arcs $[x, y] \in A^-;$
> $i := 0;$
>   **Do**
> $i := i + 1;$
>   **For** all $y \in V, t = 0, \ldots, T$ **do** $d_a(y, t)^i := d_a(y, t)^{i-1};$

Case 1: i is an odd number:
    For $t = 1, ..., T$ **do**
        For every $y \in V \backslash \{s\}$ **do** *forward searching operation:*

$$d_a(y,t)^i := \min\{sign(l(y,t-1))(d_a(y,t-1)^i + c(y,t-1)), d_a(y,t)^i,$$

$$\min_{\{x|(x,y)\in A^+\}} \min_{\{u|u+b(x,y,u)=t \wedge l(x,y,u)>0\}} \{d_a(x,u)^i + c(x,y,u)\}\};$$

Case 2: i is an even number:
    For $t = T - 1, ..., 0$ **do**
        For every $y \in V \backslash \{\rho\}$ **do** *backward searching operation:*

$$d_a(y,t)^i := \min\{sign(l[y,t+1])(d_a(y,t+1)^i + c[y,t+1]), d_a(y,t)^i,$$

$$\min_{\{x|(x,y)\in A^-\}} \min_{\{u|u+b[x,y,u]=t \wedge l[x,y,u]>0\}} \{d_a(x,u)^i + c[x,y,u]\}\};$$

While there exists at least one $d_a(y,t)^i \neq d_a(y,t)^{i-1}$;
Let $d_a^*(\rho) := \min_{0 \leq t \leq T} d_a(\rho,t)^i$;
End

<u>Lemma 4.3</u> *When the procedure SDFP-AW terminates, $d_a^*(\rho)$ is the length of a shortest dynamic f-augmenting path from s to ρ with time at most T.*

Proof: Similar to the proof of Lemma 4.1, here we only need to prove that, for each i and t, $d_a(y,t)^i$ obtained by the procedure is the length of a shortest dynamic f-augmenting path from s to y of time at most t with the alternating number at most i. Also, we only need to consider dynamic paths whose first sections are positive.

The proof is conducted by double inductions on i and t. Consider $i = 1$. Use the second induction on time t. When $t = 0$, since $i = 1$ is an odd number, no positive dynamic path $P(s,y)$ of time at most t exists in N except when $y = s$. Since there are no negative cycles in N, we know that the length of the shortest dynamic path $P(s,s)$ of time t is 0 for any t. In the initialization of procedure SDFP-AW, we have $d_a(y,0)^1 = +\infty$ ($y \in V \backslash s$) and $d_a(s,0)^1 = 0$; so the claim holds.

Assume $t > 0$ and for all values $t' < t$, $d_a(y,t')^1$ is the length of a shortest dynamic path $P(s,y)$ of time at most t' with the alternating number at most 1 for all vertices y.

Consider a vertex y. If $y = s$ then the proof is straightforward. So assume $y \neq s$. First we prove that there exists a path of time at most t with the alternating number 1 and with length $d_a(y,t)^1$. If $d_a(y,t)^1 = +\infty$, there is nothing to prove. So assume $d_a(y,t)^1$ is finite. Then by the forward searching operation, $d_a(y,t)^1$ must come from $d_a(y,t-1)^1 + c(y,t-1)$ while $l(y,t-1) > 0$, or $d_a(x,u)^1 + c(x,y,u)$ for some x such that $(x,y) \in A^+$ and some u such that $u + b(x,y,u) = t$. If the first case occurs, by the induction on t, there is a dynamic path $P(s,y)$ of time at most $t-1$ with length $d_a(y,t-1)^1$. Of course, $P(s,y)$ is also a path of time at most t with the length $d_a(y,t-1)^1 + c(y,t-1)$ since $l(y,t-1) > 0$. If the second case occurs, by the induction on t, we know that there is a path $P'(s = x_1, ..., x_{r-1} = x)$ from s to x of time u with the alternating number at most 1 and with length $d_a(x,u)^1$. We extend the path with vertex y, obtaining a path $P(s,y)$. The time of $P(s,y)$ is at most t and the length is $d_a(x,u)^1 + c(x,y,u) = d_a(y,t)^1$.

We now prove that $d_a(y,t)^1$ is the length of a shortest path from s to y of time at most t. Let $P(s = x_1, ..., x_r = y)$ be a shortest path from s to y of time at most t, and $w(x_i)$ the waiting time at x_i $(i = 1, ..., r)$. Let x be the predecessor of y on this path. Let u be the departure time of the subpath $P(s,x)$ of $P(s,y)$ at vertex x, and let $\zeta(x)$ be the length of $P(s,x)$. The definition $t = u + w(y) + b(x,y,u)$ implies $u < t$ since $b(x,y,u) > 0$ and $w(y) \geq 0$. Thus, by induction, $\zeta(x) + \sum_{\tau=\alpha(x)}^{u-1} c(x,\tau) \geq d_a(x,u)^1$. By definition, the length of $P(s,y)$ is $\zeta(y) = \zeta(x) + \sum_{\tau=\alpha(x)}^{u-1} c(x,\tau) + c(x,y,u) + \sum_{\tau=\alpha(y)}^{t-1} c(y,\tau) \geq d_a(x,u)^1 + c(x,y,u) + \sum_{\tau=\alpha(y)}^{t-1} c(y,\tau) \geq d_a(y,t)^1$. Thus, we must have $\zeta(y) = d_a(y,t)^1$, since $P(s,y)$ is a path of shortest length and since there exists a path achieving $d_a(y,t)^1$, as we showed above. This completes the proof on $i = 1$.

Assume that for $i < k$, the claim is true. Now consider $i = k$.

Suppose i is an odd number. Consider $t = 0$. Since there are no negative cycles in N, the length of the shortest dynamic path $P(s,s)$ of time 0 is 0. For $y \neq s$, since no flow can depart from y at time 0, no residual network can contain negative arcs which allow a flow to reach y from any other vertex x at time 0. This means there exists no path $P(s,y)$ of time 0. On the other hand, in the procedure, we let $d_a(y,0)^i = d_a(y,0)^{i-1}$ at first. By the induction on i we know $d_a(y,0)^{i-1} = +\infty$. By the formula of the forward searching operation, $d_a(y,0)^i$ is unchanged during the iterations since no path $P(s,x)$ with negative arrival time u in N or in any residual networks can be extended by adding an arc $(x,y) \in A^+$ with a positive $b(x,y,u)$ such that $u + b(x,y,u) = 0$. Therefore the claim is correct.

Assume the claim holds for $t' < t$. Now consider the case at time t. One may clearly see that there exists a path of time at most t with the alternating number at most i and with the length $d_a(y,t)^i$. We now prove that $d_a(y,t)^i$ is the length of a shortest path from s to y of time exactly t.

Let $P(s = x_1, ..., x_r = y)$ be a shortest path from s to y of time at most t, and $w(x_i)$ the waiting time at x_i ($i = 1, ..., r$). Let x be the predecessor of y on this path. Let u be the departure time of the subpath $P(s,x)$ of $P(s,y)$ at vertex x, and let $\zeta(x)$ be the length of $P(s,x)$. The definition $t = u + w(y) + b(x,y,u)$ implies $u < t$ since $b(x,y,u) > 0$ and $w(y) \geq 0$. Thus, by induction, $\zeta(x) \geq d_a(x,u)^i$. By definition, the length of $P(s,y)$ is $\zeta(y) = \zeta(x) + \sum_{\tau=\alpha(x)}^{u-1} c(x,\tau) + c(x,y,u) + \sum_{\tau=\alpha(y)}^{t-1} c(y,\tau) \geq d_a(x,u)^i + c(x,y,u) + \sum_{\tau=\alpha(y)}^{t-1} c(y,\tau)$, hence $\zeta(y) \geq d_a(x,u)^i + c(x,y,u) + \sum_{\tau=\alpha(y)}^{t-1} c(y,\tau) \geq d_a(y,t)^i$ according to the formula. Thus, we must have $\zeta(y) = d_a(y,t)^i$, since $P(s,y)$ is a path of shortest length and since there exists a path achieving $d_a(y,t)^i$, as we showed above.

The proof for the case with i being an even number is similar. □

Lemma 4.4 *The procedure SDFP-AW can be implemented in $O(mnT^2)$ time.*

Proof: The time needed for the initialization and the sorting are bounded by $O(nT)$ and $O(mT)$. From the description of the iterative formula, the number of iterations for i is proportional to $T \sum_x \sum_{y,(y,x)\in A} 1 = mT$. Since $i \leq nT$, the total time is bounded by $O(mnT^2)$. □

The following is the algorithm to solve the TVMCF problem with arbitrary waiting times.

> **Algorithm TVMCF-AW**
> **Begin**
> $\bar{v} := 0$;
>   **For** $j = 1, ..., v$ **do**
>   **Call** procedure SDFP-AW;
>   **If** $d^*(\rho) < +\infty$ then call the procedure UPNET; (there is an f-augmenting path $P_j(s,\rho)$ with flow value $f_j = Cap(P_j(s,\rho))$; so update the network)
>   **Else** stop; (no feasible solution to send all flow value v from s to ρ within time T)
> $\bar{v} := \bar{v} + f_j$;
>   **If** $\bar{v} \geq v$ then stop;
> **End**

Theorem 4.3 *Algorithm TVMCF-AW solves optimally the TVMCF problem with waiting times unconstrained at any vertices.*

The proof is similar to that for Theorem 4.1. By Lemma 4.4 and Algorithm TVMCF-AW, we have the following theorem.

Theorem 4.4 *The time complexity of Algorithm TVMCF-AW is bounded above by* $O(vmnT^2)$. □

4.3 Waiting at a vertex is constrained by an upper bound

In this section, we consider the TVMCF problem where waiting time at a vertex is constrained by a vertex-dependent upper bound. As we have mentioned in Section 5, Chapter 3, Theorem 3.3 gives us a feasible condition to determine whether a dynamic f-augmenting path is feasible under the bounded waiting time constraint. This condition will also be used in our algorithm to find the shortest f-augmenting path in a dynamic residual network.

Similar to Section 4.1, we will propose a procedure, SDFP-BW, to solve a subproblem as follows:

Subproblem SP3 – *Given a network N which has nonzero transit times, arbitrary costs, and no negative cycles, find a shortest feasible dynamic f-augmenting path from s to ρ within time T where waiting time at each vertex x is constrained by an upper bound.*

We need the following notation when we solve this subproblem.

Definition 4.6 *Define $d_b(x, t)^k$ as the length of a shortest feasible dynamic f-augmenting path from s to the vertex x of time exactly t with the alternating number at most k, where the waiting time at x is zero.*

Definition 4.7 *Let $P(s, x)$ be a dynamic path and (x, y) (or $[x, y]$) the next appended arc. Let u_A and u_D be the arrival time and the departure time at x, respectively. Define:*

$$\mathcal{L}_P(x, y, u_A, u_D) = \begin{cases} l(x, y, u_D) \prod_{\tau=u_A}^{u_D-1} l(x, \tau) & u_D \geq u_A \\ l(x, y, u_D) \prod_{\tau=u_A}^{u_D+1} l[x, \tau] & u_D < u_A \end{cases}$$

as the derived capacity if $(x, y) \in A^+$, or

$$\mathcal{L}_N(x, y, u_A, u_D) = \begin{cases} l[x, y, u_D] \prod_{\tau=u_A}^{u_D-1} l(x, \tau) & u_D \geq u_A \\ l[x, y, u_D] \prod_{\tau=u_A}^{u_D+1} l[x, \tau] & u_D < u_A \end{cases}$$

if $[x, y] \in A^-$.

Clearly, $\mathcal{L}_P(x, y, u_A, u_D)$ (or $\mathcal{L}_N(x, y, u_A, u_D)$) indicates the arc capacity $l(x, y, u_D)$ as well as the vertex capacity during the waiting time at vertex x. When $\mathcal{L}_P(x, y, u_A, u_D) > 0$ (or $\mathcal{L}_N(x, y, u_A, u_D) > 0$), it means that we can append the arc (x, y) (or the arc $[x, y]$) to the path $P(s, x)$ to obtain a feasible dynamic path from s to y.

Definition 4.8 *Let $P(s, x)$ be a dynamic path and (x, y) (or $[x, y]$) the next appended arc. Suppose z is the predecessor of x on P. Let*

$$\delta_\alpha(x) = \begin{cases} -r_1\{z, x, \tau(z)\} - r_2\{x, y, u_D\} + u_x & if \ \alpha_{i^0}(x) \ does \ not \ exist \\ \alpha_{i^0}(x) - u_A + u_x & otherwise. \end{cases}$$

Define

$$\mathcal{F}(x, y; t, r_1, r_2) =$$

$$\begin{cases} \{(u_A, u_D) | u_D + b(x, y, u_D) = t \wedge 0 \le u_D \le u_A + \delta_\alpha(x)\} & (x, y) \in A^+ \\ \{(u_A, u_D) | u_D + b[x, y, u_D] = t \wedge 0 \le u_D \le u_A + \delta_\alpha(x)\} & [x, y] \in A^- \end{cases}$$

as the feasible time region of x.

Definition 4.9 *Define $J(x; u_A, u_D)$ as the cost function combining the cost of the path $P(s, x)$ with the waiting cost at x:*

$$J(x; u_A, u_D) = \begin{cases} d_b(x, u_A)^i + \sum_{\tau=u_A}^{u_D - 1} c(x, \tau) & u_A \le u_D \\ d_b(x, u_A)^i + \sum_{\tau=u_A}^{u_D+1} c[x, \tau] & u_A > u_D. \end{cases}$$

Similar to the two procedures as described in Sections 4.1 and 4.2, here our idea is to use two searching operations, a *forward* pass and a *backward* pass, to calculate $d_b(y, t)^k$. Note that in the feasible condition, we need r_1 to determine $\mathcal{F}$. Consequently, at any iteration, we use a variable $\mathcal{R}_1(y, t)$ to record the current value of r_1, so that it can be brought to the next iteration to compute the new value for r_1. Moreover, we use a function, $\text{FUNA}(x, t)$, to calculate $\alpha_{i^0}(x)$ for the vertex x with respect to the arrival time t, and then use $\alpha_{i^0}(x)$ to determine $\delta_\alpha(x)$. The function $\text{FUNA}(x, t)$ is to be determined as follows:

> **FUNA**(x, t);
> **Begin**
>   **Let** $\zeta_2 := t - r_2$;
>   **Do**
>   **Let** $\zeta_1 := \zeta_2$;
>   **Let** $\zeta_2 := \max_{t_1 = \zeta_1 + 1, \dots, \zeta_1 + u_x, t_1 \le T} \alpha_1(x, t_1)$;
>   **While** $(\zeta_2 > \zeta_1) and (\zeta_2 + u_x \le T)$;
>   **If** $\zeta_1 = t - r_2$ **then Return**(-1) (it means that $a_{i^0}(x)$ does not exist);
>   **If** $\zeta_2 \le \zeta_1$ **then Return**(ζ_1) **else Return**$(T - u_x)$;
> **End.**

The procedure SDFP-BW is now described below.

Procedure SDFP-BW
 Begin
  Initialize: $d_b(s,0)^0 := 0, d_b(s,t)^0 := +\infty, t = 1,...,T;$
$d_b(y,t)^0 := +\infty, \forall y \in V\backslash\{s\}; t = 0,...,T; \; \mathcal{R}_1(y,t) := 0, \forall y \in V;$
$t = 0,...,T;$
  $i := 0;$
  Do
  $i := i + 1;$
  For all $y \in V, t = 0,\ldots,T$ **do** $d_b(y,t)^i := d_b(y,t)^{i-1};$
  Case 1: i is an odd number:
  For $t = 1,...,T$ **do**
   For every $y \in V\backslash\{s\}$ **do** *forward searching operation:*

$$d_b(y,t)^i := \min\{d_b(y,t)^i,$$

$$\min_{\{x|(x,y)\in A^+\}} \min_{\substack{(u_A,u_D)\in\mathcal{F}(x,y;t,r_1,r_2)\ and \\ \mathcal{L}_P(x,y,u_A,u_D)>0}} \{J(x;u_A,u_D) + c(x,y,u_D)\}\};$$

  Case 2: i is an even number:
  For $t = T-1,...,0$ **do**
   For every $y \in V\backslash\{\rho\}$ **do** *backward searching operation:*

$$d_b(y,t)^i := \min\{d_b(y,t)^i,$$

$$\min_{\{x|(x,y)\in A^-\}} \min_{\substack{(u_A,u_D)\in\mathcal{F}(x,y;t,\mathcal{R}_1,r_2)\ and \\ \mathcal{L}_N(x,y,u_A,u_D)>0}} \{J(x;u_A,u_D) + c[x,y,u_D]\}\};$$

  If $d_b(y,t)^i$ is determined by the second term **then**
   Let $\mathcal{R}_1(y,t) := r_1[x,y,u_D];$
  While there exists at least one $d_b(y,t)^i \neq d_b(y,t)^{i-1};$
  Let $d_b^*(\rho) := \min_{0\leq t\leq T} d_b(\rho,t)^{nT};$
 End

We have the following result on the procedure SDFP-BW.

Lemma 4.5 *When the procedure SDFP-BW terminates, $d_b^*(\rho)$ obtained is the length of a shortest feasible dynamic f-augmenting path from s to ρ with time at most T.*

Proof: Similar to Lemma 4.1, we only need to prove that for each i and t, $d_b(y,t)^i$ obtained by the procedure is the length of a shortest feasible dynamic f-augmenting path from s to y of time exactly t with the alternating number at most i. Furthermore, it will suffice for us to consider only those dynamic paths whose first sections are positive.

 The proof is carried out by double inductions on i and t. Consider $i = 1$. Use the second induction on time t. When $t = 0$, since $i = 1$ is an odd number, no positive dynamic path $P(s, y)$ of time exactly t exists in N except when $y = s$. The empty path $P(s, s)$ is feasible since the upper bounds on waiting times are assumed nonnegative. In the initialization of SDFP-BW, we have $d_b(y, 0)^1 = +\infty$ $(y \in V \backslash s)$ and $d_b(s, 0)^1 = 0$; so the claim holds.

Assume $t > 0$ and for all values $t' < t$, $d_b(y, t')^1$ is the length of a shortest feasible path $P(s, y)$ of time exactly t' with the alternating number at most 1 for all vertices y.

Consider a vertex y. First we prove that there exists a path of time exactly t with the alternating number 1 and with length $d_b(y, t)^1$. If $d_b(y, t)^1 = +\infty$, there is nothing to prove. So assume $d_b(y, t)^1$ is finite. Then by the forward searching operation, $d_b(y, t)^1$ comes from $J(x; u_A, u_D) + c(x, y, u_D)$ for some x such that $(x, y) \in A^+$ and some $(u_A, u_D) \in \mathcal{F}(x, y; t, r_1, r_2)$ while $\mathcal{L}_P(x, y, u_A, u_D) > 0$. Since $i = 1$ and $r_1 = r_2 = 0$, we have $u_A < u_D < t$. By the induction on t, we know that there is a feasible path $P'(s = x_1, ..., x_{r-1} = x)$ of time exactly u_A with the alternating number at most 1 and with length $d_b(x, u_A)^1$. We let $u_D - u_A$ be the new waiting time of P' at x. On the other hand, since $(u_A, u_D) \in \mathcal{F}(x, y; t, r_1, r_2)$ while $\mathcal{L}_P(x, y, u_A, u_D) > 0$, by Lemma 5.1, the new path is again feasible. We extend the path with vertex y, obtaining a path P with the given waiting times, and with waiting time zero at y. The time of P, with all these waiting times, is exactly t, since $u_D + b(x, y, u_D) = t$, which is the arrival time at y. The length of P is $J(x; u_A, u_D) + c(x, y, u_D) = d_b(y, t)^1$. This proves the claim.

We now prove that $d_b(y, t)^1$ is the length of a shortest feasible path of time exactly t. Let $P(s = x_1, ..., x_r = y)$ be a shortest feasible path of time exactly t with the alternating number 1. Let $w(x_i)$ be the waiting time at x_i $(i = 1, ..., r)$. So we have $w(x_r) = 0$. Let x be the predecessor of y on this path. Let u_D be the departure time of the subpath $P(s, x)$ of $P(s, y)$ at vertex x, let $u_A = u_D - w(x)$ be the arrival time at x along $P(s, x)$ and let $\zeta(x)$ be the length of $P(s, x)$. Note that $t = u_D + b(x, y, u_D)$. By induction, $\zeta(x) \geq d_b(x, u_D)^1$. By definition, the length of $P(s, y)$ is $\zeta(x) + \sum_{\tau=u_A}^{u_D-1} c(x, \tau) + c(x, y, u_D) \geq J(x; u_A, u_D) + c(x, y, u_D) \geq d_b(y, t)^1$, where the last inequality comes from the formula on the computation of $d_b(y, t)^1$. This length must be equal to $d_b(y, t)^1$ since $P(s, y)$ is a path of the shortest possible length and since there exists a path that achieves $d_b(y, t)^1$, as we showed above. This completes the proof on $i = 1$.

Assume that for $i < k$, the claim is true. Now consider $i = k$.

Suppose i is an odd number. Consider $t = 0$. Since there are no negative cycles in N, the length of the shortest feasible dynamic path $P(s, s)$ of time exactly 0 is 0. For vertex $y \neq s$, since no flow can depart from y at time 0, no residual network can contain negative arcs which allow a flow to reach y from any other vertex x at time 0. This means there exist no feasible dynamic paths $P(s, y)$ of time exactly 0. On the other hand, in the procedure, we let $d_b(y, 0)^i = d_b(y, 0)^{i-1}$ at first. By the induction on i we know $d_b(y, 0)^{i-1} = +\infty$. By the formula of the forward searching operation, $d_b(y, 0)^i$ is unchanged during the iterations since no path $P(s, x)$ with negative arrival time u_A in N or any residual networks can be extended by adding an arc $(x, y) \in A^+$ with a positive transit time $b(x, y, u_D)$ such that $u_D + b(x, y, u_D) = 0$. Therefore the claim is correct.

Assume the claim holds for $t' < t$. Now consider the case at time t. First we also prove that there exists a path of time exactly t with the alternating number at most i and with the length $d_b(y, t)^i$. If $d_b(y, t)^i = +\infty$, there is nothing to prove. So assume $d_b(y, t)^i$ is finite. Then by the forward searching operation, $d_b(y, t)^i$ comes from $d_b(y, t)^{i-1}$ or $J(x; u_A, u_D) + c(x, y, u_D)$ for some x such that $(x, y) \in A^+$ and some $(u_A, u_D) \in \mathcal{F}(x, y; t, r_1, r_2)$ while $\mathcal{L}_P(x, y, u_A, u_D) > 0$ such that $u_D + b(x, y, u_D) = t$. If the first case occurs, namely, $d_b(y, t)^i = d_b(y, t)^{i-1}$, by the induction on i, we know that $d_b(y, t)^{i-1}$ is the length of a shortest feasible dynamic path $P(s, x)$ of time exactly t with the alternating number at most $i-1$. Obviously, $P(s, x)$ is also a feasible dynamic path of time exactly t with the alternating number at most i. If the second case occurs, namely, $d_b(y, t)^i = J(x; u_A, u_D) + c(x, y, u_D)$ for some x and some u_D such that $(x, y) \in A^+$ and $(u_A, u_D) \in \mathcal{F}(x, y; t, r_1, r_2)$ while $\mathcal{L}_P(x, y, u_A, u_D) > 0$, we consider u_A. Without loss of generality, suppose that y is the jth visited vertex at time t during the ith iteration, and for any vertex which is visited before y at time t, there exists a feasible path of time exactly t with the alternating number at most i. If $u_A < t$, by the induction on t, we know that there exists a feasible path from s to x of time exactly t with the alternating number at most i. If $u_A > t$, then there must have $d_b(x, u_A)^i = d_b(x, u_A)^{i'}$ with $i' < i$. By the induction on i, we know that there exists a feasible path from s to x of time exactly t with the alternating number at most i', of course, with the alternating number at most $i - 1$. Now examine the case in where $u_A = t$. If x is visited before y at time t, by the assumption, the claim is true. Otherwise, if x has not been visited at time t, then there must have $d_b(x, u_A)^i = d_b(x, u_A)^{i''}$ with $i'' < i$. By the induction on i, we know that there exists a feasible path from s to x of time exactly t with the alternating number at most i'', of course, with the alternating number at most $i - 1$. Therefore, we can extend the path with vertex y, obtaining a path $P(s, y)$. Since $(u_A, u_D) \in \mathcal{F}(x, y; t, r_1, r_2)$ while $\mathcal{L}_P(x, y, u_A, u_D) > 0$, by Theorem 3.4, $P(s, y)$ is again feasible. The time of $P(s, y)$ is exactly t, the alternating number is at most i, and the length is $J(x; u_A, u_D) + c(x, y, u_D) = d_b(y, t)^i$. This proves the claim.

We now prove that $d_b(y, t)^i$ is the length of a shortest feasible path from s to y of time exactly t with zero waiting time at y. Let $P(s = x_1, ..., x_r = y)$ be a shortest feasible path of time exactly t with the alternating number at most i. Let x be the predecessor of y on this path. Let u_D be the departure time of the subpath $P(s, x)$ of $P(s, y)$ at vertex x, let $u_A = u_D - w(x)$ be the arrival time at x along $P(s, x)$ and let $\zeta(x)$ be the length of $P(s, x)$. By definition $t = u_D + b(x, y, u_D)$. By induction, $\zeta(x) \geq d_b(x, u_A)^i$. Clearly, the length of $P(s, y)$ is $\zeta(x) + \sum_{\tau=u_A}^{u_D-1} c(x, \tau) + c(x, y, u_D) \geq J(x; u_A, u_D) + c(x, y, u_D) \geq d_b(y, t)^i$, where the last inequality comes from the formula for the computation of $d_b(y, t)^i$. This length must be equal to $d_b(y, t)^i$ since P is a path of the shortest possible length and since there exists a path that achieves $d_b(y, t)^i$, as we showed above. This completes the proof.

The proof in the case where i is an even number can be carried out in a similar way by using induction on k (let $t = T - k$, $k = 0, ..., T$). In summary, the lemma is proved. $\square$

In the procedure SDFP-BW, to obtain $J(x; u_A, u_D)$, $\mathcal{L}_P$ and $\mathcal{L}_N$, we need to compute $\sum_{\tau=u_A}^{u_D-1} c(x, \tau)$, $\sum_{\tau=u_A}^{u_D+1} c[x, \tau]$, $\prod_{\tau=u_A}^{u_D-1} l(x, \tau)$ and $\prod_{\tau=u_A}^{u_D+1} l[x, \tau]$. Note that all these can be computed, in $O(nT^2)$ time, before the procedure SDFP-BW is applied.

Lemma 4.6 *The procedure SDFP-BW can be implemented in $O(nT^2(m + nT))$ time.*

Proof: We can see that the computing times needed for the initialization and the sorting are bounded by $O(nT)$ and $O(mT)$, respectively.

For $\mathcal{F}(x, y; t, r_1, r_2)$, note that u_D can be found in $O(1)$ time for any given t, since we have sorted them already. For condition $0 \leq u_D \leq u_A + \delta_\alpha(x)$, i.e., $\max\{0, u_D - \delta_\alpha(x)\} \leq u_A \leq T$, we can set up a set of variables, denoted by $D(x, t)$, $\forall x \in V$ and $t = 0, ..., T$, which keeps the minimal value among all $J(x; u_A, t)$ which satisfies $t - \delta_\alpha(x) \leq u_A \leq T$ and $\prod_{\tau=u_A}^{t-1} l(x, \tau) > 0$ if $u_A \leq t$ (or $\prod_{\tau=u_A}^{t+1} l[x, \tau] > 0$ if $t < u_A$). To maintain $D(x, t)$, we need to compute $\alpha_{i^0}(x)$ and $\delta_\alpha(x)$ by using function FUNA(x, u_A) after $d_b(x, u_A)^i$ is obtained. And then, for all $t \leq u_A + \delta_\alpha(x)$, calculate $J(x; u_A, t)$. If $J(x; u_A, t) < D(x, t)$ and $\prod_{\tau=u_A}^{t-1} l(x, \tau) > 0$ when $u_A \leq t$ (or $\prod_{\tau=u_A}^{t+1} l[x, \tau] > 0$ when $t < u_A$), then let $D(x, t) = J(x; u_A, t)$, otherwise, keep the current value of $D(x, t)$ unchanged. This step can be implemented in $O(T)$ time. Since the procedure has to perform this step for $t = 1, ..., T$ and for all vertices $x \in V$, it takes $O(nT^2)$ time. Therefore, for each i, the iteration can be implemented in $O(T(m + nT))$.

Since $i \leq nT$, it follows that the overall running time of this procedure is bounded above by $O(nT^2(m + nT))$.   □

The following is the algorithm to solve the TVMCF problem with bounded waiting times at vertices.

> **Algorithm TVMCF-BW**
> **Begin**
> $\bar{v} := 0$;
>   Sort all values $u + b(x, y, u)$ for $1 \leq u \leq T$ and for all arcs $(x, y) \in A^+$;
>   Sort all values $u + b[x, y, u]$ for $0 \leq u \leq T - 1$ and for all arcs $[x, y] \in A^-$;
>   Calculate $\sum_{\tau=u_A}^{u_D-1} c(x, \tau)$ and $\sum_{\tau=u_A}^{u_D+1} c[x, \tau]$ for any $x \in V$, $0 \leq u_A, u_D \leq T$;
>   Calculate $\prod_{\tau=u_A}^{u_D-1} l(x, \tau)$ and $\prod_{\tau=u_A}^{u_D+1} l[x, \tau]$ for any $x \in V$, $0 \leq u_A, u_D \leq T$;
>   **For** $j = 1, ..., v$ **do**
>   **Call** procedure SDFP-BW;
>   **If** $d^*(\rho) < +\infty$ **then** call the revised procedure UPNET; (there is an f-augmenting path $P_j(s, \rho)$ with flow value $f_j = Cap(P_j(s, \rho))$, so update the network)
>   **Else** stop; (no feasible solution to send all flow value v from s to ρ within time T)
> $\bar{v} := \bar{v} + f_j$;
>   **If** $\bar{v} \geq v$ then stop;
> **End**

Theorem 4.5 *The algorithm TVMCF-BW solves optimally the TVMCF problem with the bounded waiting time constraint at each vertex.*

The proof is similar to that for Theorem 4.1. By Lemma 4.6 and the Algorithm TVMCF-BW, we have the following theorem.

Theorem 4.6 *The running time required by the Algorithm TVMCF-BW is bounded above by $O(vnT^2(m + nT))$.*   □

5.   How to fine-tune the algorithms in special cases?

The time requirements of the algorithms we have presented in the previous sections are only worst-case upper bounds. In practice, the actual time requirements of the algorithms may be much less than this bound. In fact, we can show that in certain cases, even the worst-case upper bounds may be improved.

For the Algorithm TVMCF-ZW, we have the following result:

Corollary 4.1 *If the arc capacities and the transit costs are positive constants and the transit times are nondecreasing functions of time t, the running time of Algorithm TVMCF-ZW reduces to $O(vmnT)$.*

Proof: Under the condition of the corollary, any shortest dynamic f-augmenting path with no artificial arcs (we call this a *true shortest path* (TSP)) will not visit a vertex more than once. Otherwise, suppose $P(s = x_1, ..., x_i, x_{i+1}, ..., x_i, x_j, ..., x_r = \rho)$ is a TSP which visits the vertex x_i twice. Then we can create a new path $P'(s = x_1, ..., x_i, x_j, ..., x_r = \rho)$ by deleting the section $(x_{i+1}, ..., x_i)$ from P. It is clear that P' is a feasible path with $\zeta(P') < \zeta(P)$, since P' traverses less arcs as compared to P, while the transit times are nondecreasing functions and the transit costs are constant. This contradicts the fact that P is a shortest path.

Consequently, one can see that the first k shortest f-augmenting paths obtained by the algorithm will not visit any vertex more than k times. (Otherwise, if one represents these k paths by k TSPs, then there must exist a TSP that visits the vertex more than once.) Now let g_k be the number of iterations to search for the kth shortest f-augmenting path P_k, r_k the alternating number of P_k, and h_k the number of vertices of P_k. Clearly, $g_k = r_k + 1 \leq h_k$. Hence, we have $\sum_{k=1}^{v} g_k = \sum_{k=1}^{v}(r_k + 1) \leq \sum_{k=1}^{v} h_k \leq nv$. Noting that $O(mT)$ time is needed in each iteration, we obtain the corollary. $\square$

Similarly, the upper bound in Theorem 4.4 can also be reduced in certain cases. For example, we can have the following result:

Corollary 4.2 *If for each vertex, its capacity is unlimited, and its waiting cost at time t is less than the transit cost of each arc leaving it at time t, then the running time of Algorithm TVMCF-AW reduces to $O(vmnT)$.* $\square$

6. The time-varying maximum (k, c)-flow problem

As an application of TVMCF, we now discuss the *time-varying maximum (k, c)-flow problem*, which can be stated as follows:

Consider a time-varying network $N(V, A, b, c_p, l)$, where V is the vertex set, A is the arc set, $b(x, y, t)$ is the transit time to traverse the arc (x, y) at time t, $l(x, y, t)$ is the arc capacity, which is given but can be exceeded at a penalty of $c_p(x, y, t)$ per unit flow, where $c_p(x, y, t)$ is a nonnegative integer. Each unit of flow sent from s to ρ yields a payoff of k, where k is a positive integer. Given a time limit T, the problem is to determine how much extra arc capacity to purchase and how much *flow to be sent from the source vertex s to the sink vertex ρ within the time limit T, so as to maximize the net profit.*

The static version of the maximum (k,c)-flow problem, where both the arc capacity and the penalty are time independent, is discussed by Wagner and Wan (1993). The maximum (k,c)-flow problem has applications in transportation systems, electronic supply systems, and manpower systems.

Obviously, when $k = 1$ and all penalties $c_p(x,y,t) = 1$, the time-varying maximum (k,c)-flow problem reduces to the time-varying maximum flow problem, since any dynamic path in N has at least one arc and paying a penalty to exceed the capacity of an arc will not cause any extra net profit. This implies that the time-varying maximum (k,c)-flow problem is NP-hard too. In what follows, we will develop a pseudo-polynomial algorithm to solve the problem. We will only consider the case where waiting at a vertex is not allowed. The problem under other waiting time constraints can be solved in a similar manner.

The key idea of our method can be stated as follows. First, for each arc (x,y) in the given network N, we create a parallel arc $(x,y)^r$. We call this the *reserve arc*, and the original arc (x,y) the *normal arc*. Let the transit cost $c(x,y,t) = 0$ for the normal arc and $c(x,y,t) = c_p(x,y,t)$ for the reserve arc. Then, we apply the procedure SDFP-ZW (see Section 4.1) to find the shortest dynamic f-augmenting path in the new network. Note that, since all normal arcs have zero cost, they will be considered first in generating the shortest dynamic f-augmenting path. After the normal arcs are exhausted, those reserve arcs are then considered. In particular, we will follow the following steps:

(1) Create a new network $N'(V, A, b, c, l')$ by the following method: For each arc $(x,y) \in A$, create a reserve arc $(x,y)^r$. Let $l'(x,y,t) = l(x,y,t)$ and $c(x,y,t) = 0$ for each normal arc, and $l'(x,y,t) = \infty$ and $c(x,y,t) = c_p(c,y,t)$ for each reserve arc. Furthermore, create artificial arcs for each normal arc and reserve arc respectively, and set l and c same as those in Section 2.

(2) Find a shortest dynamic f-augmenting path P in N' by applying the SDFP-ZW procedure.

(3) Let $\zeta(P)$ be the cost of P. If $k - \zeta(P) > 0$, then that means we can earn $k - \zeta(P)$ units of profit by sending one unit of extra flow from s to ρ. Go to step (4); Otherwise, if $k - \zeta(P) \leq 0$ (that means we cannot obtain any more profit in the network), then stop.

(4) Let f_p be the flow value which we send along the path P. If $f_p = \infty$, i.e., $Cap(P) = \infty$ (that means we can send flow as much as we want

in the network, i.e., the original problem is unbounded), then stop; Otherwise, use the procedure UPNET to update the network N' based on the flow f_p (See Section 2, Chapter 3; Note that now we need update the parameters for both the normal and the reserve arcs). Still denote the residual network as N'.

(5) Let F be the total flow value and $\zeta(F)$ be the total benefit of flow F. Let $F := F + f_p$, and $\zeta(F) = \zeta(F) + f_p(k - \zeta(P))$ (initially, let $F = \zeta(F) = 0$). Go to step (2).

We can now describe our algorithm as follows.

Algorithm TVMKCF
  Begin
  Create a network $N' = (V, A, b, c, l')$ for the given network N;
  Let $F := \zeta(F) := 0$;
  Do
      Call Procedure SDFP-ZW to obtain a shortest dynamic f-augmenting path P;
      If $k - \zeta(P) > 0$ **then**
         If $Cap(P) < \infty$ **then** $F := F + f_p$, $\zeta(F) := \zeta(F) + f_p(k - \zeta(P))$, and create the dynamic residual network based on f_p;
         Else Stop (the original problem is unbounded);
      While $k - \zeta(P) > 0$;
  End.

Theorem 4.7 *When the Algorithm TVMKCF terminates, it finds either a path with $Cap(P) = \infty$ (the time-varying maximum (k, c)-flow problem is unbounded), or the optimal solution with the flow value F and the total benefit $\zeta(F)$.*

Proof: Straightforward.

We now analyze the time complexity of the algorithm. First, we can show that, if the problem is bounded, the total flow value which can be sent from s to ρ is less than or equal to UmT, where U is the maximum capacity of the normal arcs. Note that any maximum (k, c)-flow with $F < \infty$ in N can be decomposed into several subflows f_i such that f_i travels on a dynamic path P_i $(1 \le i \le l)$, where l is a finite integer, and each path P_i must contain at least one normal arc (x, y). Otherwise, suppose that there exist a path P_{i_0}, $1 \le i_0 \le l$, which consists of reserve arcs only. If $k - \zeta(P_{i_0}) \ge 0$, then the problem is unbounded since all capacities of the reserve arcs are infinite, and we can send flow as much as possible along the path P_{i_0} to get a positive profit. This contradicts the assumption of $F < \infty$. If $k - \zeta(P_{i_0}) < 0$, then deleting subflow f_{i_0}

from F can improve the total net profit. This contradicts the assumption that F is a maximum (k,c)-flow. Since the total capacity of the normal arcs is less than or equal to UmT, the claim is true. From Lemma 5.2, the procedure SDFP-ZW needs $O(mnT^2)$ running time, and at least one unit of flow can be sent from s to ρ after a dynamic f-augmenting path P is found. Therefore the total running time of the algorithm is bounded by $O(Unm^2T^2)$. Now, we consider the case that the problem is unbounded. There must exist a dynamic f-augmenting path P with $k - \zeta(P) \geq 0$. Clearly, P can be found in $UmT + 1$ steps, regardless of whether P consists of reserve arcs only or not. In summary, we have

Theorem 4.8 *Algorithm can solve the time-varying maximum (k,c)-flow problem in $O(Unm^2T^2)$ time, if waiting at any vertex is not allowed.*

7.   Additional references and comments

Aronson (1989); Aronson (1986); Klingman et al (1982) and Orlin (1984) develop solution approaches based on the multi-period structure of a dynamic network. An inductive out-of-kilter algorithm is constructed by White et al (1969) for solving a dynamic, acyclic network, formulated from the problem of distributing empty freight cars in a railroad system. Klingman et al (1982) address a multi-period production/distribution problem formulated as a minimum cost dynamic flow model. However, the structure of the network and its parameters, such as costs, capacities, and transit times, are all fixed constants. Only the node requirements may change over time. Orlin (1984) considers a problem defined over an infinite time horizon with the objective to minimize the average convex cost per period. Some variants of the problem have also been addressed. Aronson (1989) and Aronson (1986) propose to solve the problem by a forward algorithm, a procedure for dealing with multi-period problems through solving successively subproblems over longer time horizon. Carey et al (1993) study a time-varying minimum cost flow problem in which the arc cost depends on the flow transmitted on the arc.

Chapter 5

TIME-VARYING MAXIMUM CAPACITY PATH PROBLEMS

1. Introduction

The *maximum capacity path (MCP)* problem is to find a path between two vertices such that the capacity of the path is maximized, where the capacity of a path is defined as the minimum of the capacities of the arcs and vertices on this path. The problem is also called the *max-min path problem*. When all parameters are assumed to be static constants, the problem has been studied in the literature and shown to be polynomially solvable (see Gabow (1985) and Punuen (1991)). It has also been used to tackle a number of network problems. For example, Hansen (1980) has utilized MCP to find paths under bicriteria; Lawler (1976) has shown that MCP has applications in reliability theory; Berman et al (1987) has utilized MCP to obtain the optimal minimax path of a single service unit in a network; Ichimori et al (1979) has also used MCP to solve minimax flow problems.

In this chapter, we will study the *time-varying maximum capacity path (TVMCP)* problem, where the problem parameters may change over time. Specifically, we consider the situations where a flow must take a transit time $b(x, y, t)$ to traverse an arc (x, y), and both the transit time $b(x, y, t)$ and the capacity $l(x, y, t)$ of the arc (x, y) are functions of the departure time t at the vertex x. Waiting at the vertex x is allowed, subject to a waiting time constraint as well as a vertex capacity constraint $l(x, t)$, which limits the maximum flow at x during the time period $[t, t+1)$. The parameter $l(x, t)$ may also change over time t. The problem is to determine the maximum capacity path from the source s to another pre-specified vertex, subject to the constraint that the total travel time of the path is not greater than a given time limit T.

Applications of the TVMCP problem vary. The following is an example: A manufacturer is to determine a route to transport his product from his factory in one city to his customer in another city. The product is perishable, and thus it is imperative to deliver it to the customer within the time limit T. Besides, due to the high setup cost and other practical restrictions, it is not desirable to separate the transportation of the product to different routes. To make the maximum benefit, the concern of the manufacturer is to determine a single route that has the maximum transmitting capacity. As the transmitting capacity and the transit time of each arc connecting two cities are time dependent, the problem can be formulated as a TVMCP model.

We will show, in this chapter, that while the static MCP problem is polynomially solvable (Gabow (1985) and Punuen (1991)), the TVMCP problem is NP-complete. Pseudo-polynomial algorithms will be developed to find the optimal solutions for the TVMCP problem in a number of situations. The rest of this chapter is organized as follows. The complexity of the problem is studied in Section 2. Solution algorithms are developed in Section 3. A fully polynomial approximation approach is given in Section 4. Finally, some additional references and comments are given in Section 5.

2.   NP-completeness

We first show that TVMCP with no waiting allowed at any vertex is NP-complete. We denote this variant of the problem as TVMCP-NW. The decision version of TVMCP-NW can be stated as: Given a time-varying network N, a time limit T, and an integer k, does there exist a path P from s to $x = \rho$ within time T such that $Cap(P) \geq k$, where $Cap(P)$ denotes the capacity of the path P? Since no waiting is allowed at any vertex, we do not need to consider the vertex capacities now. In other words, $Cap(P)$ is now equal to the minimum capacity value among all arc capacities on the path P.

Theorem 5.1 *TVMCP-NW is NP-complete, even if the underlying graph of the network is a planar graph.*

Proof: We will show that the following Knapsack problem is reducible to TVMCP-NW:

Knapsack problem (KP): Given a set of positive integers $w_1, w_2, ..., w_n$ and B, does there exist a subset $S \subset \{1, 2, ..., n\}$ such that $\sum_{i \in S} w_i = B$?

Given any instance of KP, we can construct accordingly an instance of TVMCP-NW as follows: The network N is as shown in Figure 5.1;

$x_0 = s$ and $x_{n+1} = \rho$; $T = B + n + 1$, $k = 2$ and

$$b(x_{i-1}, x_i', t) = w_i,$$
$$b(x_i', x_i, t) = b(x_{i-1}, x_i, t) = b(x_n, x_{n+1}, t) = 1,$$
$$\text{for } 0 \le t \le T, 1 \le i \le n,$$
$$l(x_{i-1}, x_i', t) = l(x_i', x_i, t) = l(x_{i-1}, x_i, t) = 2,$$
$$\text{for } 0 \le t \le T, 1 \le i \le n,$$
$$l(x_n, x_{n+1}, t) = 0, \text{ for } 0 \le t < B + n,$$
$$l(x_n, x_{n+1}, B + n) = 2.$$

We now show that KP has a 'yes' answer iff TVMCP-NW has a 'yes' answer.

If there exists a set $S \subset \{1, 2, ..., n\}$ such that $\sum_{i \in S} w_i = B$, a path P with no waiting time at any vertex can be constructed in such a manner: Starting from x_0, we choose the arcs (x_{i-1}, x_i') and (x_i', x_i) if $i \in S$, or choose the arc (x_{i-1}, x_i) if $i \notin S$; At last, we choose the arc (x_n, x_{n+1}). Obviously, $\alpha(x_n) = B + n$. Since $l(x_n, x_{n+1}, B + n) = 2$ and $b(x_n, x_{n+1}, B + n) = 1$, we have $\alpha(x_{n+1}) = B + n + 1 \le T$ and $Cap(P) = 2$, which is the maximum possible capacity in N. Thus, there is a path p in N that achieves the maximum capacity within time T.

Figure 5.1. The network constructed for TVMCP-NW

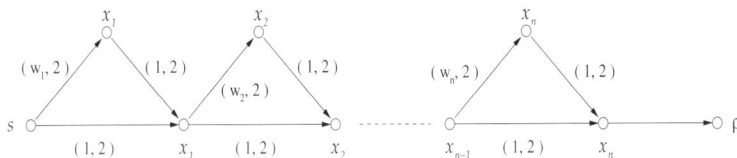

On the other hand, if there exists a path P with time T and capacity $Cap(P) = 2$, then $\alpha(x_n) = B + n$ since only when $t = B + n$, $l(x_n, x_{n+1}, t) = 2$, and the total time to traverse P is equal to $B + n + 1$. Note that the arc (x_{i-1}, x_i') takes time $b(x_{i-1}, x_i', t) = w_i$ if it is on P. Letting $S = \{i | (x_{i-1}, x_i') \in A(P)\}$, where $A(P)$ is the set of all arcs on P, we have $\sum_{i \in S} w_i = B$.

In summary, we complete the proof. □

We now consider the TVMCP problem where waiting time at each vertex x is subject to an upper bound u_x. Denote this variant of the problem as TVMCP-BW. It is clear that TVMCP-NW is a special case of TVMCP-BW with $u_x = 0$ for all $x \in V$. This gives us immediately the following result.

Theorem 5.2 *TVMCP-BW is NP-complete, even if the underlying graph of the network is a planar graph.*   □

We can also show that the TVMCP problem with arbitrary waiting time at any vertex is NP-complete. Denote the problem by TVMCP-AW. Similar to the proof for TVMCP-NW, we can show that KP is reducible to TVMCP-AW, by constructing an instance of TVMCP-AW as in Figure 6.1 and letting $l(x,t) = 1$ for all x and t. Obviously, if a dynamic path has a waiting time at a vertex, its capacity will not exceed 1In other words, if there is a path with capacity 2, it can not have any waiting times. This leads to the following result.

Theorem 5.3 *TVMCP-AW is NP-complete, even if the underlying graph of the network is a planar graph.*

3.   Algorithms

In what follows, we will give a pseudopolynomial approach which can obtain the optimal solution for the problem TVMCP-BW, where the waiting time at vertex y is bounded above by a given number u_y. The problem can be restated as: *Given a time-varying network $N(V, A, b, l)$ and a time limit T, find the maximum capacity path from s to $x \in V \backslash \{s\}$ within the time T subject to the constraint that the waiting time at any intermediate vertex y is not greater than u_y.*

Definition 5.2 *Let $\xi_b(x,t)$ be the maximum capacity of the path from s to x of time exactly t and with waiting time zero at x, subject to the constraint that waiting time at any vertex y on the path is not greater than u_y. If such a path does not exist, let $\xi_b(x,t) = 0$.*

Lemma 5.1 *$\xi_b(s,0) = \infty$, $\xi_b(s,t) = 0$ for $t = 1, ..., T$, and $\xi_b(y,0) = 0$ for all $y \neq s$. For $t > 0$, we have:*

$$\xi_b(y,t) = \max_{(x,y)\in A} \max_{(u_A,u_D)\in\mathcal{F}(x,y,t)} \{\min\{\xi_b(x,u_A), \mathcal{L}(x,u_A,u_D), l(x,y,u_D)\}\}$$

where $\mathcal{F}(x,y,t) = \{(u_A,u_D)|u_D + b(x,y,u_D) = t, 0 \leq u_D - u_A \leq u_x\}$, $u_A = \alpha(x)$, $u_D = \tau(x)$, and $\mathcal{L}(x,u_A,u_D) = \min_{u_A \leq t' \leq u_D-1} l(x,t')$.

Proof: The case $t = 0$ is obvious. There does not exist any path of time zero from s to any other vertex y since all transit times $b(x,y,t)$ are positive. Note that the empty path $P(s,s)$ is feasible. Hence $\xi_b(s,0) = \infty$ and $\xi_b(y,0) = 0$ for all $y \neq s$.

We now use induction to prove the formula. Consider $t = 1$. If there exists a path from s to y of time exactly one, then y must be a neighbor of s with $b(s,y,0) = 1$. The formula holds with $u_A = u_D = 0$ and $x = s$.

Assume that the claim is true for all $t' < t$. Now, examine the case where time is t. Consider a vertex y.

If $\xi_b(y, t) = 0$, there is nothing to prove. So assume $\xi_b(y, t) > 0$. First, we show that there exists a path from s to y of time exactly t and with capacity $\xi_b(y, t)$. By the formula, $\xi_b(y, t) = \min\{\xi_b(x, u_A), \mathcal{L}(x, u_A, u_D), l(x, y, u_D)\}$ for some x such that $(x, y) \in A$ and some $(u_A, u_D) \in \mathcal{F}(x, y, t)$. By induction, we know that there is a feasible path, $P'(s, x)$, from s to x of time exactly u_A and with capacity $\xi_b(x, u_A)$. Letting $u_D - u_A$ be the new waiting time at x and extending $P'(s, x)$ with vertex y, we can obtain a path $P(s, y)$ of time exactly t. Since $0 \leq u_D - u_A \leq u_x$ and $\mathcal{L}(x, u_A, u_D) > 0$, path $P(s, y)$ is again feasible. The capacity of P is $\min\{\xi_b(x, u_A), \mathcal{L}(x, u_A, u_D), l(x, y, u_D)\} = \xi_z(y, t)$. The claim is therefore proven.

We now prove that $\xi_b(y, t)$ is the capacity of a maximum capacity path from s to y of time exactly t. Let $P(s = x_1, x_2, ..., x_r = y)$ be a maximum capacity feasible path from s to y of time exactly t, x be the predecessor of y on P, u_A and u_D be the arrival time and departure time of the subpath P' from s to x at vertex x, respectively. By definition, $t = u_D + b(x, y, u_D)$. By induction, $Cap(P') \leq \xi_b(x, u_A)$ since $u_D < t$. By definition, $Cap(P) = \min\{Cap(P'), \mathcal{L}(x, u_A, u_D), l(x, y, u_D)\} \leq \min\{\xi_b(x, u_A), \mathcal{L}(x, u_A, u_D), l(x, y, u_D)\} \leq \xi_b(y, t)$, where the last inequality comes from the computation of the formula. This capacity must equal $\xi_b(y, t)$, since P is a path with the maximum possible capacity and of time exactly t, and since there exists a path of time exactly t that achieves the capacity $\xi_b(y, t)$, as we showed above. This completes the proof. $\qquad\square$

To improve the efficiency of the algorithm to be presented below, we will use a binary heap to store the values of $\xi_b(x, u_A)$ for every vertex x and for all $max\{0, t - u_A\} \leq u_A \leq t$. Particularly, given (x, y) and t, a value of u_D that satisfies $u_D + b(x, y, u_D) = t$ is known. Thus, the corresponding value of $l_a(x, y, u_D)$ is also known. Therefore, the problem of finding

$$\max_{(u_A, u_D) \in \mathcal{F}(x, y, t)} \{\min\{\xi_b(x, u_A), \mathcal{L}(x, u_A, u_D), l(x, y, u_D)\}\}$$

reduces to solving a problem of finding

$$\xi_b^m(x, u_D) = \max_{max\{0, u_D - u_x\} \leq u_A \leq u_D} \{\min\{\xi_b(x, u_A), \mathcal{L}(x, u_A, u_D)\}\}$$

(recall the definition of $\mathcal{F}(x, y, t)$ in Lemma 5.1). In the procedure TVMCP-BW below, we keep all maximal value $\xi_b^m(x, u)$, for all x and all $u \leq t - 1$. And for each vertex x, maintain one heap $Heap_x$. After $\xi_b(x, t)$ is obtained, the new $Heap_x$ at time t is obtained by deleting

$\xi_b(x, t-u_x-1)$ from the heap (if $x-u_x-1 \geq 0$) and inserting $\xi_b(x,t)$. Let $\xi_b^*(x, t-1)$ be the maximum value in $Heap_x$ after deleting $\xi_b(x, t-u_x-1)$ and before inserting $\xi_b(x,t)$. Then, we have

Lemma 5.2 *For $t = 0$, $\xi_b^m(x,t) = 0$. For $t = 1, .., T$,*

$$\xi_b^m(x,t) = \begin{cases} \max\{\min\{\xi_b^*(x,t-1), l(x,t-1)\}, \xi_b(x,t)\} \\ \qquad\qquad\qquad \text{if } \xi_b^*(x,t-1) < \xi_b^m(x,t-1) \\ \max\{\min\{\xi_b^m(x,t-1), l(x,t-1)\}, \xi_b(x,t)\} \\ \qquad\qquad\qquad \text{if } \xi_b^*(x,t-1) \geq \xi_b^m(x,t-1) \end{cases}$$

Proof: It is obvious to see $\xi_b^m(x,0) = 0$. Thus, we need only prove the formula for $t > 0$. By the definition,

$$\xi_b^m(x,t) = \max_{\max\{0,t-u_x\} \leq t' \leq t} \{\min\{\xi_b(x,t'), \mathcal{L}(x,t',t)\}\}$$

$$= \max\{\min\{\max_{\max\{0,t-u_x\} \leq t' \leq t-1} \{\min\{\xi_b(x,t'), \mathcal{L}(x,t',t-1)\}\},$$

$$l(x,t-1)\}, \xi_b(x,t)\} \qquad (5.1)$$

and

$$\xi_b^m(x,t-1) = \max_{\max\{0,t-u_x-1\} \leq t' \leq t-1} \{\min\{\xi_b(x,t'), \mathcal{L}(x,t',t-1)\}\}.$$

Now we consider two cases:

- Case 1: $\xi_b^*(x,t-1) \geq \xi_b^m(x,t-1)$. If $t \leq u_x$, then any elements will not be deleted in $Heap_x$ at time t, and

$$\xi_b^m(x,t-1) = \max_{0 \leq t' \leq t-1} \{\min\{\xi_b(x,t'), \mathcal{L}(x,t',t-1)\}\}.$$

Thus,

$$\xi_b^m(x,t) = \max\{\min\{\xi_b^m(x,t-1), l(x,t-1)\}, \xi_b(x,t)\}.$$

If $t > u_x$, then $t - u_x - 1 \geq 0$. Therefore,

$$\xi_b^m(x,t-1) = \max_{t-u_x-1 \leq t' \leq t-1} \{\min\{\xi_b(x,t'), \mathcal{L}(x,t',t-1)\}\}$$

$$= \max\{\min\{\xi_b(x,t-u_x-1), \mathcal{L}(x,t-u_x-1,t-1)\},$$

$$\max_{t-u_x \leq t' \leq t-1} \{\min\{\xi_b(x,t'), \mathcal{L}(x,t',t-1)\}\}\}. \qquad (5.2)$$

By the assumption, $\xi_b^*(x,t-1) \geq \xi_b^m(x,t-1)$, and by the definition, $\mathcal{L}(x, t-u_x-1, t-1) \leq \max_{t-u_x \leq t' \leq t-1} \mathcal{L}(x,t',t-1)$. These imply

$$\xi_b(x,t-u_x-1) \leq \max_{t-u_x \leq t' \leq t-1} \xi_b(x,t').$$

Thus,

$$\xi_b^m(x, t-1) = \max_{t-u_x \le t' \le t-1} \{\min\{\xi_b(x, t'), \mathcal{L}(x, t', t-1)\}\}.$$

By substituting $\xi_b^m(x, t-1)$ for $\max_{t-u_x \le t' \le t-1}\{\min\{\xi_b(x, t'), \mathcal{L}(x, t', t-1)\}\}$ in (5.1),we can see that the formula holds.

- Case 2: $\xi_b^*(x, t-1) < \xi_b^m(x, t-1)$. It implies $t > u_x$. We can now examine (5.2). Since

$$\max_{t-u_x \le t' \le t-1} \xi_b(x, t') \ge \max_{t-u_x \le t' \le t-1}\{\min\{\xi_b(x, t'), \mathcal{L}(x, t', t-1)\}\},$$

we have

$$\max_{t-u_x \le t' \le t-1} \xi_b(x, t') = \xi_b^*(x, t-1)$$

$$< \xi_b^m(x, t-1)$$

$$= \min\{\xi_b(x, t-u_x-1), \mathcal{L}(x, t-u_x-1, t-1)\}$$

$$\le \mathcal{L}(x, t-u_x-1, t-1)$$

$$\le \min_{t-u_x \le t' \le t-1} \mathcal{L}(x, t', t-1)$$

That is, $\max_{t-u_x \le t' \le t-1}\xi_b(x, t') < \min_{t-u_x \le t' \le t-1}\mathcal{L}(x, t', t-1)$. Thus,

$$\xi_b^*(x, t-1) = \max_{t-u_x \le t' \le t-1} \xi_b(x, t')$$

$$= \max_{t-u_x \le t' \le t-1} \{\min\{\xi_b(x, t'), \mathcal{L}(x, t', t-1)\}\}.$$

By substituting $\xi_b^*(x, t-1)$ for $\max_{t-u_x \le t' \le t-1}\{\min\{\xi_b(x, t'), \mathcal{L}(x, t', t-1)\}\}$ in (5.1), we can obtain

$$\xi_b^m(x, t) = \max\{\min\{\xi_b^*(x, t-1), l(x, t-1)\}, \xi_b(x, t)\}.$$

This completes the proof. □

The following procedure calculates $\xi_b(y, t)$.

Procedure TVMCP-BW;
Initialize $\xi_b(s, 0) := \infty$, $\xi_b(s, t) = 0$ for $t = 1, ..., T$; $\xi_b(y, 0) := 0$ for all $y \ne s$ and $t = 0, 1, ..., T$; $Heap_x := \{\xi_b(x, 0)\}$ and $\xi_b^m(x, 0) := \xi_b(x, 0)$ for all x;
Sort all values $u + b(x, y, u)$ for all $u = 1, ..., T$ and for all arcs $(x, y) \in A$;
For $t = 1, 2, ..., T$ **do**
For all $y \in V \backslash \{s\}$ **do**

$$\xi_b(y,t) := \max_{(x,y)\in A} \max_{\{u_D|u_D+b(x,y,u_D)=t\}} \{\min\{\xi_b^m(x,u_D), l(x,y,u_D)\}\};$$

For every vertex y update $Heap_y$ as follows
  If $t > u_y$ **then delete-heap**$_y \xi_b(y, t-u_y-1)$;
  Let $\xi_b^*(y, t-1) := $ **Maximum-heap**$_y$;
  If $\xi_b^*(y, t-1) \le \xi_b^m(y, t-1)$ **then**

$$\xi_b^m(y,t) := \max\{\min\{\xi_b^*(y,t-1), l(y,t-1)\}, \xi_b(y,t)\};$$

Else

$$\xi_b^m(y,t) := \max\{\min\{\xi_b^m(y,t-1), l(y,t-1)\}, \xi_b(y,t)\};$$

  Insert-heap$_y \xi_b(y,t)$;
Return;

Our main algorithm can be stated below.

  Algorithm TVMCP-BW Begin
    Call procedure TVMCP-BW;
    For every y **do** $\xi_b(y) := \max_{0 \le t \le T} \xi_b(y,t)$;
  End.

Lemma 5.3 *After Algorithm TVMCP-BW terminates, $\xi_b(y)$ is the capacity of a maximum capacity path from s to y of time at most T.*

Lemma 5.3 comes directly from Lemma 5.1, and therefore its proof is omitted here.

Lemma 5.4 *Algorithm TVMCP-BW can be implemented such that it runs in $O(T(m + n \log T))$ time.*

Proof: It is obvious that the initialization can be done in $O(Tn)$ time. The sorting can be implemented in $O(Tm)$ time when we use bucket-sort. The two steps of inserting $\xi_b(y,t)$ and deleting $\xi_b(y, t-u_y-1)$ take $O(\log u_y) = O(\log T)$ time. The step of calculating $\xi_b^m(y,t)$ takes constant time. Since the procedure TVMCP-BW has to perform these two steps for all $t = 1, 2, ..., T$ and all vertex $y \in V$, it takes in total $O(Tn \log T)$ time to maintain the heaps. The step of finding $\xi_b^m(y,t)$ takes $O(1)$ time. Finally, the last step of computing $\xi_b(y)$ for all $y \in V$ takes $O(Tn)$ time. Therefore the overall running time of the algorithm is bounded above by $O(T(m + n \log T))$. □

Combining Lemmas 5.3-5.4, we have

Theorem 5.4 *The TVMCP-BW problem can be optimally solved in $O(T(m + n \log T))$ time.*

For the problem TVMCP-NW, since no waiting is allowed at any vertex, we need not employ any heap to store the vertex capacities $l(x, t)$ between time $u_D - u_x \leq t \leq u_D$. Thus, the algorithm TVMCP-BW can be easily applied to this case. This is given below.

Definition 5.3 *Let $\xi_z(x, t)$ be the maximum capacity of the path from s to x of time exactly t, where no waiting is allowed at any vertex. If such a path does not exist, let $\xi_z(x, t) = 0$.*

Lemma 5.5 $\xi_z(s, 0) = \infty$, $\xi_z(s, t) = 0$ *for* $t = 0, 1, ..., T$, *and* $\xi_z(y, 0) = 0$ *for all* $y \neq s$. *For* $t > 0$, *we have:*

$$\xi_z(y, t) = \max_{(x,y) \in A} \max_{\{u | u + b(x,y,u) = t\}} \{\min\{\xi_z(x, u), l(x, y, u)\}\}.$$

The proof for the above lemma is similar to that for Lemma 5.1. Based on Lemma 5.5, we can have the following procedure to compute $\xi_z(y, t)$.

Procedure TVMCP-NW
    Initialize $\xi_z(s, 0) := \infty$, and $\xi_z(s, t) = \infty$ for $t = 1, ..., T$;
$\xi_z(y, t) := 0$ for all
$y \neq s$ and $t = 0, 1, ..., T$;
    Sort all values $u + b(x, y, u)$ for all $u = 1, ..., T$ and for all arcs
$(x, y) \in A$;
    For $t = 1, 2, ..., T$ **do**
      For all $y \in V \setminus \{s\}$ **do**

$$\xi_z(y, t) := \max_{(x,y) \in A} \max_{\{u | u + b(x,y,u) = t\}} \{\min\{\xi_z(x, u), l(x, y, u)\}\}$$

    Return;

The following is the algorithm to solve the problem TVMCP-NW:

Algorithm TVMCP-NW
    Begin
      Call procedure TVMCP-NW;
      For every vertex y **do** $\xi_z(y) := \max_{0 \leq t \leq T} \xi_z(y, t)$;
    End.

Lemma 5.6 *After Algorithm TVMCP-NW terminates, $\xi_z(y)$ is the capacity of a maximum capacity path from s to y of time at most T, where no waiting is allowed at any vertex.*

Proof: The correctness of the procedure TVMCP-NW follows directly from Lemma 5.5. Since $\xi_z(\rho, t)$ is the capacity of the maximum capacity path from s to ρ of time exactly t, where $t \in [0, T]$, the maximum of

$\xi_z(\rho, t)$ over all $0 \leq t \leq T$ must be the capacity of the maximum capacity path from s to ρ within time T. Moreover, the path generated by the algorithm contains no waiting time at any vertices. $\qquad\square$

Lemma 5.7 *Algorithm TVMCP-NW can be implemented such that it runs in $O(T(m+n))$ time.*

Proof: It is obvious that the initialization step needs $O(Tn)$ time. For the sorting step, we can use bucketsort, with T buckets. Since there are Tm values to be sorted, this step can be performed in $O(Tm)$ time.

Since the values $u + b(x, y, u)$ are now sorted, the overall time needed to compute $\xi_z(y, t)$ is proportional to $T \sum_y \sum_{x,(x,y) \in A} 1 = Tm$. Thus, the total running time of the algorithm is bounded by $O(T(m+n))$. $\square$

In summary, we obtain

Theorem 5.5 *The TVMCP-NW problem can be optimally solved in $O(T(m+n))$ time.*

We now consider the problem TVMCP-AW, where $u_x = \infty$ for each vertex $x \in V$. In other words, waiting at any vertex is arbitrarily allowed. In this case, a path from s to x of time at most t is also a path of time at most $t + 1$, as long as the capacity of the vertex allows us to wait at x during the period $[t, t+1]$.

Definition 5.4 *Let $\xi_a(x, t)$ be the maximum capacity of the path from s to x of time at most t, where waiting at any vertex is not limited. If such a path does not exist, let $\xi_a(x, t) = 0$.*

Similar to Lemma 5.5, we can have:

Lemma 5.8 $\xi_a(s, t) = \infty$, *for $t = 0, 1, ..., T$, and $\xi_a(y, 0) = 0$ for all $y \neq s$. For $t > 0$,*

$$\xi_a(y, t) = \max_{(x,y) \in A} \max_{\{u \mid u + b(x,y,u)=t\}} \{\min\{\xi_a(y, t-1), l(y, t-1)\},$$

$$\min\{\xi_a(x, u), l(x, y, u)\}\}.$$

We can also use the following procedure to compute $\xi_a(x, t)$.

> **Procedure TVMCP-AW;**
> **Initialize** $\xi_a(s, t) := \infty$ for $t = 0, 1, ..., T$; $\xi_a(y, 0) := 0$ for all $y \neq s$ and $t = 0, 1, ..., T$;
> **Sort** all values $u + b(x, y, u)$ for all $u = 1, ..., T$ and for all arcs $(x, y) \in A$;
> **For** $t = 1, 2, ..., T$ **do**

For all $y \in V \backslash \{s\}$ **do**

$$\xi_a(y,t) := \max_{(x,y) \in A} \max_{\{u | u + b(x,y,u) = t\}} \{\min\{\xi_a(y,t-1), l(y,t-1)\},$$

$$\min\{\xi_a(x,u), l(x,y,u)\}\}$$

Return;

The algorithm solves the problem TVMCP-AW.

> **Algorithm TVMCP-AW**
> **Begin**
>   **Call** procedure TVMCP-AW;
>   **For** every y **do** $\xi_a(y) := \max_{0 \le t \le T} \xi_a(y,t)$;
> **End.**

Similar to our analysis of the problems TVMCP-NW and TVMCP-BW, we can prove the following results.

Lemma 5.9 *After Algorithm TVMCP-AW terminates, $\xi_a(y)$ is the capacity of a maximum capacity path from s to y for the problem TVMCP-AW.*

Lemma 5.10 *Algorithm TVMCP-AW can be implemented such that it runs in $O(T(m+n))$ time.*

Theorem 5.6 *The problem TVMCP-AW can be optimally solved in $O(T(m+n))$ time.*

4. Finding approximate solutions

We can see that the time requirements of the algorithms developed above depend on the parameter T. In this section we will show how the time complexity of the algorithms could be reduced provided that we aim to find approximate solutions only. The key idea is to evaluate only a subset of the values for $t = \{0, 1, \cdots, T\}$. We will analyze the problem TVMCP-AW; that is, waiting at any vertex is not constrained. Furthermore, we assume that there is no vertex capacity limit; namely, $l(x,t) = \infty$ for all vertex x and all time t.

Specifically, for a given network $N(V, A, b, l)$, we will apply our algorithm TVMCP-AW to a new problem TVMCP-AW′, which is same as the original problem TVMCP-AW except that $t = 0, k, 2k, ..., k\lfloor T/k \rfloor$, and $b'(x,y,t) = k \cdot \lceil b(x,y,t)/k \rceil$ for $t = 0, k, 2k, ..., k \cdot \lfloor T/k \rfloor$. It follows from Theorem 5.6 that, Algorithm TVMCP-AW can find an optimal solution P^0 for the problem TVMCP-AW′ in a time $O(T(m+n)/k)$, which can be made sufficiently small if k is large. The question now is how to ensure the solution P^0 to be a satisfactory approximate solution when it is applied to the original problem.

Let $\xi'_a(x,t)$ be the maximum capacity of the path from s to x of time exactly t for the problem TVMCP-AW$'$. The following is an application of algorithm TVMCP-AW to determine the optimal solution for the problem TVMCP-AW$'$.

> **Algorithm TVMCP-AW$'$**
> **Begin**
>   **Initialize** $\xi'_a(s,0) := \infty$ and $\xi'_a(s,t) := 0$ for $t = k, ..., k\lfloor T/k \rfloor$;
> $\xi'_a(y,0) := 0$ for all $y \neq s$ and $t = 0, k, ..., k\lfloor T/k \rfloor$;
>   **Sort** all values $u + b'(x,y,u)$ for all $u = 0, k, ..., k\lfloor T/k \rfloor$ and for all arcs $(x,y) \in A$;
>   **For** $t = k, 2k, ..., k\lfloor T/k \rfloor$ **do**
>   **For** each $y \in V \backslash \{s\}$ **do**
>
> $$\xi'_a(y,t) := \max_{(x,y)\in A} \max_{\{u|u+b'(x,y,u)=t\}} \{\max\{\xi'_a(x,t-k),$$
>
> $$\min\{\xi'_a(x,u), l(x,y,u)\}\}\}$$
>
>   **For** every vertex y **do** $\xi'_a(y) := \max_{0 \leq t \leq k\lfloor T/k \rfloor} \xi'_a(y,t)$;
> **End.**

First, we can show the following result. For ease of reference, we use $Cap(P)$ to denote the capacity of a path P.

Lemma 5.11 *If Algorithm TVMCP-AW$'$ finds a solution P^o, then there exists a path P in the original network N such that $Cap(P^o) = Cap(P)$.*

Proof: Suppose $P^o(s = x_1, ..., x_r = \rho)$ with $\alpha^o(x_i)$, $\tau^o(x_i)$, and $w^o(x_i)$ as its arrival time, departure time, and waiting time at vertices x_i, $i = 1, ..., r$, respectively. We first show that a feasible path P can be constructed for TVMCP-AW, and then prove that $Cap(P^o) = Cap(P)$.

We can construct a path P of the same topological structure as P^o, and let $\tau(x_i) = \tau^o(x_i)$, for $i = 1, ..., r$. Furthermore, we can let $\alpha(x_i) = \tau(x_{i-1}) + b(x_{i-1}, x_i, \tau(x_{i-1}))$. Consequently, $w(x_i) = \alpha^o(x_i) - \alpha(x_i) + w^o(x_i)$. Note that we always have $\alpha^o(x_i) \geq \alpha(x_i)$ due to the definition of b'. This is valid for all i. Therefore, the path P with all $\alpha(x_i)$, $\tau(x_i)$ and $w(x_i)$ as given above comprises a feasible dynamic path in the original network, which can be traversed within the time limit T.

Since $\tau(x_i) = \tau^o(x_i)$ for all i, it is clear that $Cap(P^o) = Cap(P)$. $\square$

Let $\delta(x,y) = \max_{1 \leq t \leq T-1}\{l(x,y,t+1) - l(x,y,t)\}$ for each arc $(x,y) \in A$ and let $\delta_{max} = \max_{(x,y) \in A} \delta(x,y)$. Further, let C_{min} denote the minimum arc capacity. Define the relative error of the solution P^0 as

$$r = \frac{Cap(P^*) - Cap(P^0)}{Cap(P^*)},$$

where P^* is the optimal solution of the original problem TVMCP-AW. The following theorem indicates that P^0 is an approximate solution for TVMCP-AW.

Theorem 5.7 *For any given ε, we can choose $k = [C_{min}\varepsilon/(n\delta_{max})]$. Then, Algorithm TVMCP-AW' can find, in $O(T/k(m+n))$ time, a solution P^0 such that $r \leq \varepsilon$.*

Proof: By Lemma 5.11, the solution P^0 obtained by Algorithm TVMCP-AW' is a feasible solution of TVMCP-AW.

We now consider another solution P', which has the same topological structure as P^*, while the departure time at the beginning vertex of each arc takes a value only at $t = 0, k, 2k, ..., k\lfloor T/k \rfloor$. Then we have

$$Cap(P') \leq Cap(P^0) \leq Cap(P^*)$$

since P' is a feasible solution of TVMCP-AW' while P^0 is the optimal solution of TVMCP-AW'. It therefore follows that

$$r = \frac{Cap(P^*) - Cap(P^0)}{Cap(P^*)} \leq \frac{Cap(P^*) - Cap(P')}{Cap(P^*)} \leq \frac{nk\delta_{max}}{Cap(P^*)} \leq \frac{nk\delta_{max}}{C_{min}}.$$

If we choose $k = [C_{min}\varepsilon/(n\delta_{max})]$, then $r \leq \varepsilon$. On the other hand, it follows from Lemma 5.10 that the solution P^0 can be found in a time $O(T/k(m+n))$. □

If $l(x, y, t)$ is a linear function of t for all (x, y), then we can see that

$$\delta_{max} \leq \frac{L_{max} - C_{min}}{T},$$

where L_{max} is the maximum capacity of all $l(x, y, t)$. If we choose

$$k = \frac{TC_{min}\varepsilon}{n(L_{max} - C_{min})},$$

then we have $O(T/k(m+n)) = O(n(m+n)(L_{max} - C_{min})(\varepsilon C_{min})) = O(L_{max}n(m+n)/(\varepsilon C_{min}))$. This gives us the following result.

Corollary 5.1 *Algorithm TVMCP-AW' can be implemented in $O(L_{max}n(m+n)/(\varepsilon C_{min}))$ time, if $l(x, y, t)$ is a linear function of t for all (x, y).*

Table 5.1. $b(x, y, t)$ and $l(x, y, t)$

t	(s,g)	(s,h)	(h,g)	(h,f)	(g,f)	(g,ρ)	(f,ρ)
0	2,41	1,44	2,45	2,44	1,43	2,42	1,43
1	1,42	2,44	2,39	1,44	1,43	1,40	1,43
2	1,41	2,41	1,40	3,45	2,45	1,41	2,44
3	3,43	1,43	2,45	3,43	1,43	1,41	1,43
4	2,40	3,42	1,42	1,42	1,44	2,42	1,41
5	2,44	2,41	2,44	1,41	1,40	2,41	3,40
6	3,45	1,40	3,43	2,40	2,41	1,40	1,42
7	1,41	1,44	2,41	2,40	1,41	3,41	1,41
8	2,41	2,44	2,45	3,44	1,42	1,44	2,45
9	3,43	2,43	1,44	1,45	1,45	1,42	1,44

From Corollary 5.1, we know that Algorithm TVMCP-AW$'$ becomes fully polynomial if L_{max}/C_{min} is bounded above by a polynomial in m or n.

Now, we will give an example to illustrate how to obtain an approximate solution.

Example 5.1

Given a time-varying network as shown in Figure 5.2. All transit time and arc capacity as listed in Table 5.1 ($T = 9$).

Figure 5.2.   The network for Example 5.1

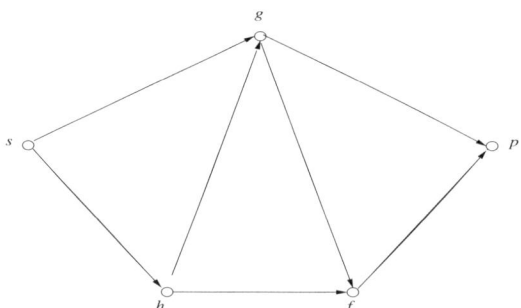

One can find the maximum capacity path $P = (s, h, f, \rho)$ with $\alpha(\rho) = 4$ and $Cap(P) = 6$ ($\tau(s) = 0$, $\alpha(h) = \tau(h) = 1$, and $\alpha(f) = \tau(f) = 2$). Now we try to use Algorithm TVMCP-AW$'$ to find an approximate

Table 5.2. $b'(x, y, t)$ and $l(x, y, t)$

t	(s, g)	(s, h)	(h, g)	(h, f)	(g, f)	(g, ρ)	(f, ρ)
0	3,41	3,44	3,45	3,44	3,43	3,42	3,43
3	3,43	3,43	3,45	3,43	3,43	3,41	3,43
6	3,45	3,40	3,43	3,40	3,41	3,40	3,42
9	3,43	3,43	3,44	3,45	3,45	3,42	3,44

solution P' with $\varepsilon = 2$. Notice that we have $C_{min} = 38$, $n = 5$, and $\delta_{max} = 5$, therefore, $k = [C_{min}\varepsilon/(n\delta_{max})] = [76/25] = 3$. Let $b' = \lceil b/3 \rceil$ and obtain a new table as shown in Table 5.2.

Then, we find a path $P' = (s, h, f, \rho)$ with $\tau(s) = 0$, $\alpha(h) = \tau(h) = 3$, $\alpha(f) = \tau(f) = 6$, $\alpha(\rho) = 9$ and $Cap(P') = 4$. Actually, in the original network, the corresponding path could be found with $\tau(s) = 0$, $\alpha(h) = 1$, $w(h) = 2$, $\tau(h) = 3$, $\alpha(f) = 6$, $\tau(f) = 6$, and $\alpha(\rho) = 7$.

5. Additional references and comments

The well-known approach of creating an equivalent static time-expanded network to model the time-varying network is also applicable to handle the TVMCP problem. The computational time for solving, for example, the TVMCP model with the bounded waiting time constraint will be $O(\min\{T(m + nT + n\log(nT)), (mT + nT^2)\log_n W\})$ in the worse case if one applies the algorithm of Gabow (1985) on the time-expanded network, where m is the number of arcs, n is the number of vertices and W is the maximum arc capacity. The computational time of the algorithm we describe in this chapter is, however, bounded by $O(T(m + n\log T))$.

The TVMCP problem may appear as a subproblem in the process of solving other time-varying network optimization problems. For example, as we have seen in Chapter 4, the time-varying maximum flow problem can be solved by repeatedly finding a dynamic augmenting path in the dynamic residual network. Theoretically, it is not necessary to require each augmenting path to have the maximum capacity. It will, however, reduce the computational requirement if each augmenting path is also a maximum capacity path (see Ahuja et al (1993)). Similarly, TVMCP can also be applied in solving the time-varying versions of the problems studied in Hansen (1980); Lawler (1976); Berman et al (1987); Ichimori et al (1979).

Chapter 6

THE QUICKEST PATH PROBLEM

1.   Introduction

Consider a network $N(V, A, l, b)$, where $G = (V, A)$ is a directed graph without multiple arcs and self loops, $l(x, y) \geq 0$ and $b(x, y) > 0$ are the capacity and the lead time (transit time) for an arc $(x, y) \in A$, respectively. If $P(x_1, x_2, ..., x_k)$ is a path in N, then the lead time of the path P is defined as $\ell(P) = \sum_{i=1}^{k-1} b(x_i, x_{i+1})$, and the capacity of the path P is $Cap(P) = \min_{1 \leq i \leq k-1} l(x_i, x_{i+1})$. To send σ units of flow from x_1 to x_k through P, one can send a batch of $Cap(P)$ units at each time $t = 0, 1, 2, \cdots$. This process can continue until all σ units of flow are transmitted, which requires a total transmission time $\ell(p) + \lceil \sigma/Cap(P) \rceil$. The *quickest path (QP) problem* is to find the path that can send the σ units of flow from s to ρ with the minimum total transmission time.

Chen and Chin (1990) have pointed out that the quickest path problem does not possess the property that "any subpath of a shortest path must itself be a shortest path". The reason is that the transmission time of a path depends not only on the lead time of the path, but also on the path capacity. They have also proposed an algorithm which can optimally solve the problem in $O(m^2 + mn \log m)$ time. Rosen, Sun and Xue (1991) have developed another algorithm which can be implemented in $O(m^2 + mn \log n)$ time.

In this chapter we consider a variant of the QP problem, which is similar to the QP version that has studied in the literature, but is more general in the objective function. Specifically, instead of considering the total transmission time $\ell(P) + \lceil \sigma/Cap(P) \rceil$, we are interested in a more general function $h(\ell(P), \wp(P))$, where $\wp(P) = \lceil \sigma/Cap(P) \rceil$, which is the number of batches to transmit the flow due to the capacity limit of the path. In other words, the total cost is a function of the lead time of the path and the number of time to divide the flow into small batches to pass through the path. Such a cost function may model many practical problems in logistics management, telecommunications, etc. For example, suppose a manufacturer has to deliver his product from his manufacturing base to a market through a transportation network, where air, sea, and road transportation modes are available but their speed, cost, and routing are different. If he decides to deliver his product by air, the transit time is shorter but his cargo will have to be divided into smaller batches due to the capacity limit of aircraft. If he decides to use sea transportation, the transit time is longer but the cargo may be transported in larger batches. A natural objective to be optimized is to determine the best solution (path) so that the total logistics cost is minimized.

We deal with the situation where the problem parameters may change over time. Specifically, we consider the model in which $l(x, y, t)$ and $b(x, y, t)$ are the capacity and the lead time of the arc (x, y) at time $t = 0, 1, ..., T$, where T is a given positive integer, $l(x, y, t)$ is a non-negative integer and $b(x, y, t)$ is a positive integer for any (x, y) and t. The problem is to determine, by taking into account the time-varying information, a path P such that $h(\ell(P), \wp(P))$ is minimized, where $h(\ell(P), \wp(P))$ is a non-decreasing function of $\wp(P)$. Note that when $h(\ell(P), \wp(P))$ takes the specific form $\ell(p) + \lceil \sigma/Cap(P) \rceil$ with all capacities and lead times being constant, our model reduces to the traditional QP problem. To be consistent with the terminology that has been used in the literature, we call such a path the *quickest path (QP)*, and our model the time-varying QP problem.

This chapter is organized as follows. Section 2 is the detailed problem formulation. Section 3 shows that the problem is NP-hard, even when the underlying graph of the network is a directed planar graph. Pseudo-polynomial algorithms are developed in Section 4. As an application, in Section 5 we examine a static *k-quickest path problem*, which is to determine the first k quickest paths for any given k. Some additional references and remarks are given in Section 6.

2.   Problem formulation

Consider a time-varying network $N(V, A, l, b)$, where $l(x, y, t) \geq 0$ and $b(x, y, t) > 0$ are the capacity and the lead time (transit time) of an arc $(x, y) \in A$ at time t. We assume that there is no capacity limit at any vertex. We further assume that the capacity $l(x, y, t)$ is a nonnegative integer and the lead time $b(x, y, t)$ is a positive integer. A single path is to be determined to send σ units of flow from the source vertex s to the sink vertex ρ, such that the total cost $h(\ell(P), \wp(P))$ (see Section 1 above) is minimized, where $h(\ell(P), \wp(P))$ is a non-decreasing function of $\wp(P)$.

Definition 6.1 *Let $P(x_1, ..., x_r)$ be a path from x_1 to x_r. The arrival time of a vertex x_i on P is defined as $\alpha(x_i)$ such that $\alpha(x_1) = t_0 \geq 0$ (for the source vertex s, we let $\alpha(s) = 0$), and*

$$\alpha(x_i) = \alpha(x_{i-1}) + w(x_{i-1}) + b(x_{i-1}, x_i, \tau(x_{i-1})), \qquad for\ i = 2, ..., r,$$

where $w(x_{i-1})$ is the waiting time at vertex x_{i-1} and $\tau(x_i)$, the departure time of a vertex x_i on P, is defined as

$$\tau(x_i) = \alpha(x_i) + w(x_i), \qquad for\ i = 1, ..., r-1.$$

Definition 6.2 *Let $P(s, x)$ be a path from s to x. The time of P is defined as $\alpha(x) + w(x)$.*

Let $P(x_1, x_2, ..., x_r)$ be a path from x_1 to x_r. Then, the lead time of P is

$$\ell(P) = \sum_{i=1}^{r-1} b(x_i, x_{i+1}, \tau(x_i)) + \sum_{i=1}^{r-1} w(x_i).$$

The capacity of the path is defined as the minimum arc capacity along the path. Because of this capacity limit, the flow σ must be divided into $\wp(P)$ batches, where $\wp(P) = \lceil \sigma / Cap(P) \rceil$. We assume that when the flow starts to be transmitted, the $\wp(P)$ batches of flow will be transmitted at the same time (in other words, we assume that the time interval between sending two batches is negligible).

Given a time duration $[0, T]$, the time-varying quickest time (TVQP) problem is to find an optimal path P^* to send the σ units of flow from the source vertex s to the sink vertex ρ, such that the cost function $h(\ell(P), \wp(P))$ is minimized while all other constraints including the capacity constraints are satisfied.

3.   NP-hardness

We will now show that the general TVQP problem is NP-hard. To simplify the description, we consider the problem with zero waiting time constraint first. Let $h(\ell(P), \wp(P)) = \ell(P) + \lceil \sigma / Cap(P) \rceil$, and rewrite it as $h(P)$. The decision version of TVQP can be stated as: Given a time-varying network N, a time limit T, and an integer k, does there exist a path P from s to $x = \rho$ within time T such that $h(P) \leq k$?

Theorem 6.1 *TVQP is NP-hard, even if the underlying graph of the network is a directed planar graph.*

Proof: We show that the Knapsack problem defined below is reducible to TVQP:

Knapsack problem (KP): Given a set of positive integers $w_1, w_2, ..., w_n$ and B, does there exist a subset $S \subset \{1, 2, ..., n\}$ such that $\sum_{i \in S} w_i = B$?

Given any instance of KP, we construct accordingly an instance of TVQP as follows: The network N is as shown in Figure 6.1; $x_0 = s$ and $x_{n+1} = \rho$; $T = B + n + 1$, $\sigma = 1$, $k = B + n + 2$ and

$$b(x_{i-1}, x'_i, t) = w_i,$$
$$b(x'_i, x_i, t) = b(x_{i-1}, x_i, t) = b(x_n, x_{n+1}, t) = 1,$$
$$\text{for } 0 \le t \le T, 1 \le i \le n,$$
$$l(x_{i-1}, x'_i, t) = l(x'_i, x_i, t) = l(x_{i-1}, x_i, t) = 1,$$
$$\text{for } 0 \le t \le T, 1 \le i \le n,$$
$$l(x_n, x_{n+1}, t) = 0, \text{ for } 0 \le t < B + n,$$
$$l(x_n, x_{n+1}, B + n) = 1.$$

We now show that KP has a 'yes' answer iff TVQP has a 'yes' answer.

Figure 6.1. The constructed network for TVQP

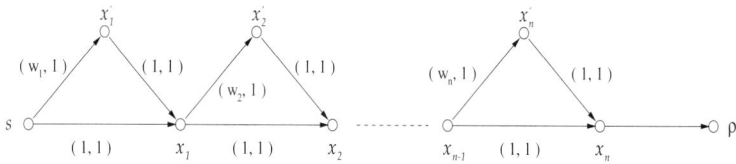

If there exists a set $S \subset \{1, 2, ..., n\}$ such that $\sum_{i \in S} w_i = B$, a path p with no waiting times at any its vertices can be constructed as follows: Starting from x_0, choose the arcs (x_{i-1}, x'_i) and (x'_i, x_i) if $i \in S$, and choose the arc (x_{i-1}, x_i) if $i \notin S$; At last, choose the arc (x_n, x_{n+1}). Obviously, $\alpha(x_n) = B + n$. Since $l(x_n, x_{n+1}, B + n) = 1$ and $b(x_n, x_{n+1}, B + n) = 1$, we have $\alpha(x_{n+1}) = B + n + 1$ and $Cap(P) = 1$. The lead time of P is $\ell(P) = \alpha(x_{n+1}) = B + n + 1$ and the cost required is $h(P) = \ell(P) + \lceil \sigma / Cap(P) \rceil = B + n + 1 + 1 \le k$. Thus, there is a path P in N that achieves the cost k.

On the other hand, if there exists a quickest path P within time duration T with $h(P) \le k$, then $\ell(P) < h(P) \le k$ as $\lceil \sigma / Cap(P) \rceil > 0$, and $\alpha(x_n)$ must be $B + n$ since only when $t = B + n$, $l(x_n, x_{n+1}, t) = 1$. Let $S = \{i | (x_{i-1}, x'_i) \in A(p)\}$, where $A(p)$ is the set of all arcs on p. Then we must have $\sum_{i \in S} w_i = B$.

In summary, we complete the proof of the theorem. $\square$

Similar results for the problems with arbitrary waiting time constraint and bonded waiting time constraint can be obtained, which are omitted here.

4.   Algorithms

Recall that the maximum capacity path problem discussed in Chapter 5 is to determine a path P from s to ρ with the maximum capacity $Cap(P)$. We can show that, the TVQP problem can be tackled by solving a set of time-varying maximum capacity path problems.

For each $t = 0, 1, ..., T - 1$, we can find a time-varying maximum capacity path $P(s, \rho)$ from s to ρ such that it can be traversed in a time exactly equal to t with $w(\rho) = 0$. Denote this path as P_t, and let

$$h^* = \min_{0 \leq t \leq T, Cap(P_t) > 0} h(\ell(P_t), \wp(P_t)).$$

We can see that h^* is the cost of the quickest path from s to ρ to transmit a given flow σ. This gives us the following property.

Property 6.1 *Given a network $N(V, A, l, b)$, a flow σ, a time limit T, and a path $P(s = x_1, x_2, ..., x_r = \rho)$ with $t = \sum_{i=1}^{r-1} b(x_i, x_{i+1}, \tau(x_i)) + \sum_{i=1}^{r-1} w(x_i) \leq T$, then*

(i) *if P is a quickest path to transmit the flow σ with the lead time $\ell(P) = t$, then there must exist a maximum capacity path of time exactly t that has the same cost as that of P, where the transit time is the lead time for each arc;*

(ii) *if P_t is a maximum capacity path of time $t \leq T$ which satisfies*

$$h(\ell(P_t), \wp(P_t)) = \min_{0 \leq t' \leq T, Cap(P_{t'}) > 0} h(\ell(P_{t'}), \wp(P_{t'}))$$

then P is a quickest path to transmit the flow σ within time duration T.

Proof: We prove part (i) first. Suppose that P is a quickest path to transmit the flow σ. If P is not a maximum capacity path of time exactly t, that is, there exists another path P' of time exactly t that has $Cap(P') > Cap(P)$, then, $h(\ell(P'), \wp(P')) \leq h(\ell(P), \wp(P))$, since $\wp(P') \leq \wp(P)$, $\ell(P') = t = \ell(P)$ and $h(\ell(P), \wp(P))$ is a non-decreasing function of $\wp(P)$. Because P is the quickest path we must have $h(\ell(P'), \wp(P')) = h(\ell(P), \wp(P))$.

We now prove part (ii). Suppose P_t is a maximum capacity path from s to ρ of time exactly t such that

$$h(\ell(P_t), \wp(P_t)) = \min_{0 \leq t' \leq T, Cap(P_{t'}) > 0} h(\ell(P_{t'}), \wp(P_{t'}))$$

Suppose P'' is the quickest path for the TVQP problem with the lead time $\ell(P'') = t''$. Let $P_{t''}$ be the maximum capacity path of time exactly t''. It follows from part (i) that

$$h(\ell(P_t), \wp(P_t)) \leq h(\ell(P_{t''}), \wp(P_{t''})) \leq h(\ell(P''), \wp(P'')).$$

Since P'' is the quickest path, we must have $h(\ell(P_t), \wp(P_t)) = h(\ell(P''), \wp(P''))$. By definition, we know that P_t is a quickest path. This completes the proof. □

Note that Property 6.1 is true for the three types of waiting time constraints. It also reveals an interesting fact that, for the case where the waiting time at a vertex is arbitrary, the optimal strategy may let the flow wait at some vertex. This contradicts a general intuition that departure of the flow at the earliest possible time would be the optimal solution.

It follows from Property 6.1 that solving the time-varying quickest path problem can be converted into solving a set of time-varying maximum capacity path problems. Let $\xi(x, t)$ be the maximum capacity of the dynamic path from s to x of time exactly t with $\alpha(x) = t$. We have the following algorithm.

> **Algorithm TVQP**
> **Begin**
>   **Call** Procedure TVMCP-ZW (TVMCP-AW, or TVMCP-BW);
>   **Let** P_t be the maximum capacity path of time exactly t
> $(t = 0, 1, ..., T)$;
>   **Let** $h^* := h := \inf$;
>   **For** $t = 0, 1, 2, ..., T$ **do**
>   **If** $\xi(\rho, t) > 0$ **then** $h := h(\ell(P_t), \wp(P_t))$;
>   **If** $h^* > h$ **then** $h^* := h$;
>   **End.**

After the algorithm terminates, h^* is the cost of a quickest path P that sends σ from s to ρ. The path P can be obtained by a backtrack procedure.

Theorem 6.2 *When Algorithm TVQP stops, h^* is the cost of a quickest path from s to ρ for a given flow σ.*

The correctness of Algorithm TVQP follows directly from Property 6.1, Lemma 5.3, Lemma 5.6, and Lemma 5.9. The following example illustrate how to use the algorithm to solve a problem with $u_x = 0$.

Example 6.1

Consider a network as given by Figure 6.2 and Table 6.1. Suppose $\sigma = 9$ and $T = 10$. The problem is to find a quickest path to send σ units of flow from s to ρ, within the time limit T. The cost function is $h(\ell(P), \wp(P)) = \ell(P) + \wp(P) = t + \lceil \sigma/\xi(\rho, t) \rceil$.

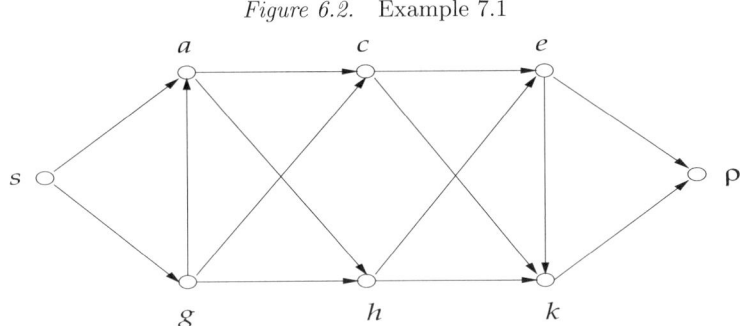

Figure 6.2. Example 7.1

Applying Algorithm TVQP, we obtain Table 6.2, which contains the maximum capacity $\xi(\rho, t)$ of the path from s to ρ with the arrival time t, and the corresponding cost $t + \lceil \sigma/\xi(\rho, t) \rceil$. It is clear that the minimum cost to transmit the flow σ is 10. By a backtrack procedure, we can find that the quickest path $P = (s, g, h, e, k, \rho)$.

The time complexity of Algorithm TVQP is dominated by the running time of the procedure TVMCP-ZW, TVMCP-AW or TVMCP-BW. Therefore, we have

Theorem 6.3 *Algorithm TVQP can optimally solve the time-varying quickest path problem with zero waiting time, arbitrary waiting time and bounded waiting time constraints in $O(T(m + n))$, $O(T(m + n))$ and $O(T(m + n \log T))$ time, respectively.*

5. The static k-quickest path problem

As an application of the time-varying quickest path problem, we will develop, in this section, a polynomial algorithm to solve the static k-quickest path problem based on the idea of replacing the lead time by the transit time for each arc $(x, y) \in A$ and setting the cost function as $h(\ell(P), \wp(P)) = \ell(P) + \wp(P) = t + \lceil \sigma/\xi(\rho, t) \rceil$.

The k-quickest path problem is to find the first, the second, ..., and the kth quickest paths from a source vertex s to the destination $x \in V \backslash \{s\}$. Rosen, Sun and Xue Rosen et al (1991) propose an method which can solve this problem in $O(rkmn + rkn^2 \log n)$ time, where r is the number

Table 6.1. $b(x, y, t)$ and $l(x, y, t)$

t	(s,a)	(s,g)	(g,a)	(a,c)	(a,h)	(g,c)	(g,h)
0	2,5	4,3	3,5	1,2	4,3	3,7	3,6
1	2,5	4,6	1,1	3,1	3,5	1,5	1,6
2	2,3	2,1	1,3	2,1	2,3	1,7	2,1
3	1,3	3,6	1,5	1,6	3,2	1,7	1,2
4	3,2	1,1	2,3	1,3	2,7	1,3	1,4
5	1,4	3,6	2,5	3,6	1,3	2,3	2,2
6	1,1	1,1	1,4	2,1	3,4	3,1	3,2
7	1,3	1,2	1,5	2,7	4,4	2,7	1,3
8	1,6	4,7	3,2	2,2	3,3	2,3	1,4
9	3,3	3,5	4,2	1,2	1,6	4,1	3,4
10	1,3	2,2	2,4	1,2	1,2	4,6	2,3

t	(c,e)	(c,k)	(h,e)	(h,k)	(e,ρ)	(e,k)	(k,ρ)
0	1,1	1,4	2,4	2,5	2,7	2,7	2,6
1	2,7	2,6	4,5	1,7	2,2	3,4	2,6
2	2,4	1,1	1,5	4,5	1,4	2,1	2,4
3	1,3	3,1	3,4	3,6	2,3	4,3	3,3
4	1,3	1,1	4,4	4,5	1,1	2,3	1,6
5	3,1	3,5	2,5	1,4	1,7	1,4	4,3
6	2,5	1,5	2,2	3,5	2,2	1,1	3,3
7	4,2	1,2	1,2	1,1	1,6	2,3	1,5
8	2,5	2,7	3,1	1,1	3,3	3,2	1,4
9	1,6	1,4	2,2	1,5	2,3	3,6	1,3
10	2,1	3,2	1,7	2,5	1,7	3,5	1,3

Table 6.2. $\xi(\rho, t)$

t	1	2	3	4	5	6	7	8	9	10
$\xi(\rho, t)$	0	0	0	0	0	1	0	3	3	3
$t + \lceil \sigma / \xi(\rho, t) \rceil$	-	-	-	-	-	14	-	10	11	12

of different arc capacities in N. Since $r \leq m$, This time complexity is bounded by $O(kmn(m + n \log n))$. The algorithm we develop below has a time complexity bounded above by $O(m(r + k)(\log n + \log r + k)) = O(m(m + k)(\log n + \log m + k))$. To illustrate, we first develop a polynomial algorithm to solve the static quickest path problem, i.e., the case $k = 1$.

The basic idea is similar to that for Algorithm TVQP. That is, we first find all maximum capacity paths from s to ρ of time exactly t under the zero waiting time constraint, and then choose the quickest one among all these paths. Let $\xi_q(y, t)$ denote the capacity of the maximum capacity path from s to y of time exactly t. For each vertex $y \in V$, set a queue H_y to contain $(\xi_q(y, t), t)$, under the following rules:

(i) All $(\xi_q(y, t), t)$ are sorted in nondecreasing order on t;

(ii) If there exist two $(\xi_q(y, t_1), t_1)$ and $(\xi_q(y, t_2), t_2)$ in H_y satisfying $\xi_q(y, t_1) \geq \xi_q(y, t_2)$ and $t_1 < t_2$, then delete $(\xi_q(y, t_2), t_2)$ from H_y.

Let us explain rule (ii). Suppose $P_1(s, y)$ and $P_2(s, y)$ are two paths from s to y with arrival times t_1 and t_2 at y, and $Cap(P_1) = \xi_q(y, t_1)$ and $Cap(P_2) = \xi_q(y, t_2)$, respectively. For a given σ, we have $h_1 = \lceil \sigma / Cap(P_1) \rceil + t_1$ and $h_2 = \lceil \sigma / Cap(P_2) \rceil + t_2$, where h_1 and h_2 are the costs of P_1 and P_2, respectively. Since $t_1 < t_2$ and $Cap(P_1) \geq Cap(P_2)$, we have $h_1 < h_2$. Thus, it is impossible to have P_2 as a quickest path and therefore $(\xi_q(y, t_2), t_2)$ can be deleted from H_y.

We also set a queue Q to store the first element in each H_y and keep the elements in Q in nondecreasing order on t. Initially, let Q contain $(\xi_q(s, 0) = \infty, 0)$. Select the first element in Q, say $(\xi_q(x, t), t)$, and maintain H_x by the rules described above. Check each arc $(x, y) \in A$ and let $\xi_q(y, t') = \min\{\xi_q(x, t), l(x, y)\}$, where $t' = t + b(x, y)$. Then, we insert $(\xi_q(y, t'), t')$ into H_y. Insert the element that is next to $(\xi_q(x, t), t)$ of H_x, if any, into Q. Then check the first element in Q again. Repeat this process until Q becomes empty. Finally, let $h = \min_{0 \leq t \leq T}\{\lceil \sigma / \xi_q(\rho, t) \rceil + t\}$, which is the cost of the quickest path. The algorithm can now be described as follows.

Algorithm QP

Begin

Initialize $Q := H_s := \{(\xi_q(s, 0) = \infty, 0)\}$; for any $x \in V \backslash \{s\}$, $H_x := \emptyset$;

While $Q \neq \emptyset$ **do**

Pick up the first element $(\xi_q(x, t), t)$ of Q and declare it as having been checked;

Check H_x. Delete any $(\xi_q(x, t^0), t^0)$ if the conditions $(\xi_q(x, t^0), t^0) \leq (\xi_q(x, t), t)$ and $t^0 > t$ are satisfied. Continue this process until we find an element that violates any of the conditions, or all $\xi_q(x, t^0)$ have been deleted;

For each arc $(x, y) \in A$ **do**

Let $t' := t + b(x, y)$;

Let $\delta(y, t') := \min\{\xi_q(x, t), l(x, y)\}$;

Do case

Case 1. $H_y = \emptyset$ or all elements in H_y have been checked. Then let $\xi_q(y, t') := \delta(y, t')$; $H_y := H_y \cup \{(\xi_q(y, t'), t')\}$ and insert $(\xi_q(y, t'), t')$ into Q;

Case 2. $H_y \neq \emptyset$. Let $t_1 := \max_{t'' \leq t', (\xi_q(y, t''), t'') \in H_y} t''$. If $t_1 < t'$ then let $\xi_q(y, t') := \delta(y, t')$; Otherwise, let $\xi_q(y, t') := \max\{\xi_q(y, t_1), \delta(y, t')\}$. Insert $(\xi_q(y, t'), t')$ in H_y. If $(\xi_q(y, t'), t')$ becomes the first element in H_y, then replace the element $(\xi_q(y, u), u)$ in Q by $(\xi_q(y, t'), t')$;

End *case*;

Remove $(\xi_q(x, t), t)$ from Q and then insert the element that is next to $(\xi_q(x, t), t)$ in H_x, if any, into Q;

End *while*;

Let $h := \min_{(\xi_q(\rho, t), t) \in H_\rho} \{\lceil \sigma/\xi_q(\rho, t) \rceil + t\}$;

End.

To illustrate, let us consider an example as follows.

Example 6.2

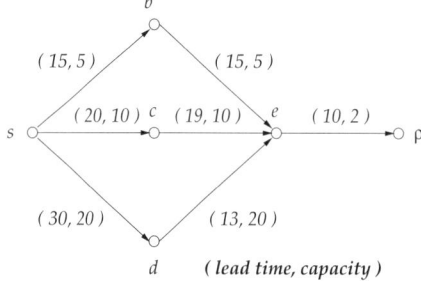

Figure 6.3. Example 6.2

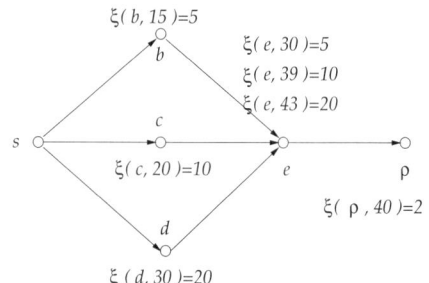

Figure 6.4. Example 6.2 (continued)

Suppose that the network is given as in Figure 6.3, where the two numbers associated with each arc are the lead time and the arc capacity, s is the source vertex, ρ is the sink vertex, and $\sigma = 100$. The problem is to find the quickest path to send all σ units of flow from s to ρ.

First, we transform the original network into a dynamic network by letting $b(x, y)$ equal the lead time of arc (x, y). Then, set $\xi(s, 0) = \infty$ and let $Q = H_s = \{(\xi(s, 0), 0)\}$. Set other $H_x = \emptyset$. Denote $(\xi(x, t), t) = (\xi(s, 0), 0)$ and delete it in Q. Consider arc (s, b) first. Because $t = 0$, we have $t' = t + b(s, b) = 0 + 15 = 15$, and $\delta(b, 15) = \min\{\xi(s, 0), l(s, b)\} = \min\{\infty, 5\} = 5$. Thus $\xi(b, 15) = 5$. Put it into H_b. Similarly, we obtain $\xi(c, 20) = 10$ and $\xi(d, 30) = 20$. Insert $(\xi(b, 15), 5)$, $(\xi(c, 20), 20)$ and $(\xi(d, 30), 20)$ into Q.

Now, we have $H_b = \{(\xi(b,15),15)\}$, $H_c = \{(\xi(c,20),20)\}$ and $H_d = \{(\xi(d,30),30)\}$. Denote $(\xi(x,t),t) = (\xi(b,15),15)$ since it is the first element in Q. Examine arc (b,e) and obtain $\xi(e,30) = 5$. Similarly, we can obtain $\xi(e,39) = 10$, and $\xi(e,43) = 20$. Finally, we obtain $\xi(\rho,40) = 2$. Since there is only one element in H_ρ, we obtain one maximum capacity path from s to ρ, which is the quickest path in the original network. The minimum cost is $h = \sigma/\xi(\rho,40) + 40 = 100/2 + 40 = 90$ (see Figure 6.4).

Theorem 6.4 *After Algorithm QP terminates, h is the transmission time of the quickest path from s to ρ.*

Proof: According to Property 6.1, we only need to prove that, when the algorithm terminates, the following statements are true for any $0 \le t \le T$ and any vertex y:

(i) If $\xi_q(y,t) \in H_y$, then $\xi_q(y,t)$ is the capacity of the maximum capacity path from s to y of time at most t,

(II) If $\xi_q(y,t) \notin H_y$, then there exists no path from s to y of time exactly t, or there exists such a path but there is another element $\xi_q(y,t')$ in H_y such that $\xi_q(y,t') \ge \xi_q(y,t)$ and $t' < t$. For the latter case, we have

$$\lceil \sigma/\xi_q(y,t')\rceil + t' \le \lceil \sigma/\xi_q(y,t)\rceil + t.$$

Thus, the path of time t' is better than that of time t.

Consider the case with $t = 0$. Since there exists no path from s to $y \in V\backslash\{s\}$ of time exactly zero except the path $P(s,s)$, we have $\xi_q(s,0) = \infty$ and $H_y = \emptyset$ for all vertices $y \ne s$. Obviously the claim is true.

Now, we use induction to prove the claim. Consider $t = 1$. If there exists a path from s to y of time exactly one, then y must be a neighbor of s with $b(s,y) = 1$. By the algorithm, $\xi_q(y,1)$ must be in H_y and it is the maximum capacity of the path from s to y of time exactly one.

Assume that for any $t' < t$, the claim is true. Now consider the case where the time is t.

First, examine the situation where $(\xi_q(y,t),t) \in H_y$. By the formula in the algorithm, $\xi_q(y,t)$ must come from $\delta(y,t)$, and there exists an element $(\xi_q(x,u),u)$ in H_x such that $(x,y) \in A$ and $u + b(x,y) = t$. Noting that $u < t$ since $b(x,y) > 0$ and $(\xi_q(x,u),u) \in H_x$, by induction, we know that there exists a maximum capacity path $P(s,x)$ from s to x of time exactly u with $\xi_q(x,u)$ as its capacity. Then, we can append arc (x,y) to $P(s,x)$, and obtain a path $P(s,y)$ from s to y with $\xi_q(y,t) = \min\{\xi_q(x,u),l(x,y)\}$ as its capacity. Now, we prove that P is the maximum capacity path from s to y of time t. Suppose that there exists another path, $P'(s, y)$, with $Cap(P') > Cap(P)$ and $t' \leq t$, where t' is the arrival time of P' at y, and x' is the predecessor of y on P'. By induction, $(\xi_q(x', u'), u')$ must be in $H_{x'}$ (any subpath of a maximum capacity path must be the maximum capacity path), where u' is the arrival time of P' at x'. Notice that $u' < t$ since $b(x', y) > 0$. By the algorithm, $\xi_q(x', u')$ should be checked earlier than $\xi_q(y, t)$. By the formula, in Case 2, $(\xi_q(y, t), t)$ should be replaced by $(Cap(P'), t')$, and $(\xi_q(y, t), t)$ can not appear in H_y. This is a contradiction.

Next, consider the case where $(\xi_q(y, t), t) \notin H_y$. Let x be the predecessor of y on P and u is its arrival time. If $(\xi_q(x, u), u) \notin H_x$, by induction, we know that either there does not exist a path from s to x of time exactly u, or the path is dominated by other paths. Therefore P is also dominated by other paths since $l(x, y)$ is a constant. Otherwise, if $(\xi_q(x, u), u) \in H_x$, then $(\xi_q(x, u), u)$ should be checked in the algorithm. By the formula in Case 2, the path is replaced by another path P'' with $t'' < t$, where t'' is the arrival time of P'' at y, and $Cap(P'') \geq Cap(P)$. Therefore the claim is also true.

In summary, we complete the proof.   □

Theorem 6.5 *Algorithm QP can be implemented in $O(rm(\log n + \log r))$ time, where r is the number of different arc capacities in N.*

Proof: The initialization step needs $O(n)$ time. Let r'_x be the number of elements in H_x ($x \in V$) and $r' = \max_{x \subset V} r'_x$. Then, finding and deleting $\xi_q(x, t)$ in Q need $O(1)$ time and inserting the next element of H_x in Q needs $O(\log n)$ time. For checking H_x and deleting redundant elements, we need $O(r'm)$ time in the whole while-loop, since there exist at most $r'n$ elements entering Q which generate at most $r'm$ elements among all queues H_y. For case 1, we need $O(\log n)$ time to insert $(\xi_q(y, t'), t')$ into Q. And for case 2, we need $O(\log r')$ time to find t_1. Since $r'_y \leq r'$, the total time for checking arcs in the whole while-loop is at most $r'm$. Hence the total running time of the while-loop is bounded by $O(r'm(\log n + \log r'))$. To compute τ it requires $O(r')$ time. Obviously, $r' \leq r$ where r is the number of different arc capacities in N, since all elements in H_x should be kept on increasing order. Therefore, the total running time of the algorithm is bounded by $O(rm(\log n + \log r))$.   □

We are now ready to tackle the k-quickest path problem. For each $y \in V \backslash \{s\}$, we first calculate the k-maximum capacity paths from s to y of time exactly t. Since we do not know which k paths will be the solution, we extend, for each vertex y, the label $\xi_q(y, t)$ to a set, namely, we let $\xi_q(y, t) = \{\xi_q^1(y, t), ..., \xi_q^k(y, t)\}$, where $\xi_q^i(y, t)$ is the capacity of the ith maximum capacity path from s to y of time exactly t and all

$\xi_q^i(y,t)$ are sorted in non-increasing order. If this path does not exist, let $\xi_q^i(y,t) = 0$. We keep k values for each $\xi_q(y,t)$, to ensure that no solutions would be lost. Before describing our algorithm, let us introduce some definitions below:

Definition 6.4 *Let $Q = \{q_1, q_2, ..., q_k\}$ and $R = \{r_1, r_2, ..., r_k\}$ be two sets with k elements and assume that all elements are sorted in nonincreasing order. Define an operation to get k minimal elements among R and Q, denoted by "mink", as follows:*

$$\text{mink}\{Q, R\} = \{k \text{ minimal elements in } Q \cup R\}.$$

Similarly, define "maxk" as follows:

$$\text{maxk}\{Q, R\} = \{k \text{ maximal elements in } Q \cup R\}.$$

Definition 6.5 *Let c be a constant and $Q = \{q_1, q_2, ..., q_k\}$ be a set with k elements. Define the division operation between c and Q, denoted by "/", as follows:*

$$c/Q = \{c/q_1, c/q_2, ..., c/q_k\}.$$

Similar to Algorithm QP, we use a queue H_x to contain all $(\xi_q(x,t), t)$ for vertex x, and keep all elements of H_x in nondecreasing order on t. Different from Algorithm QP, we maintain H_x according to the following lemma:

Lemma 6.1 *Suppose $H_x = \{(\xi_q(x,t_1), t_1), (\xi_q(x,t_2), t_2), ..., (\xi_q(x,t_f), t_f)\}$ and $\xi_q^i(x,t_j)$ is the ith value in $\xi_q(x,t_j)$ ($1 \le k$ and $1 \le j \le f$). $\xi_q^i(x,t_j)$ can be deleted in $\xi_q(x,t_j)$ if there are $k-i+1$ values in $\cup_{1 \le g \le j-1}\xi_q(x,t_g)$ which are greater than or equal to $\xi_q^i(x,t_j)$.*

Proof: First consider $\xi_q^{j'}(x,t_j)$ for $1 \le j' \le i-1$. Since $\xi_q^{j'}(x,t_j) \ge \xi_q^i(x,t_j)$, we have $\lceil \sigma/Cap(P_{j_{j'}}) \rceil + t_j \le \lceil \sigma/Cap(P_{j_i}) \rceil + t_j$, where $P_{j_{j'}}$ and P_{j_i} are the two paths with capacities $\xi_q^{j'}(x,t_j)$ and $\xi_q^i(x,t_j)$ respectively. Similarly, for those in $\cup_{1 \le g \le j-1}\xi_q(x,t_g)$ which are greater than or equal to $\xi_q^i(x,t_j)$, we can obtain the same result. This means that the path P_{j_i} cannot be the i'th quickest path ($i' \le k$) since we already have $i-1+k-i+1 = k$ paths whose transmission times are less than that of P_{j_i}. This completes the proof. □

The algorithm is given below.

Algorithm KQP
Begin
  Initialize $Q := H_s := \{(\xi_q(s,0) = \{\infty, 0, ..., 0\}, 0)\}$; for any $x \in V \backslash \{s\}$, $H_x := \emptyset$;
  While $Q \neq \emptyset$ **do**
    Pick up the first element $(\xi_q(x,t), t)$ of Q;
      Examine H_x. Delete $(\xi_q(x,t'), t')$ in H_x if the conditions $(\xi_q(x,t^0), t^0) \leq (\xi_q(x,t), t)$ and $t^0 > t$ are satisfied. Continue this process until we find an element which violates any of the conditions or all elements $(\xi_q(x,t'), t')$ have been deleted;
    For each arc $(x,y) \in A$ **do**
      Let $t' := t + b(x,y)$;
      Let $\delta(y,t') := \{\min\{\xi_q^1(x,t), l(x,y)\}, ..., \min\{\xi_q^k(x,t), l(x,y)\}\}$;
      If $\delta^i(y,t') \leq b_y^k$ $(1 \leq i \leq k)$ **then** let $\delta^i(y,t') = 0$;
      Do case
        Case 1. $H_y = \emptyset$ or all elements in H_y have been checked. Then let $\xi_q(y,t') := \delta(y,t')$; $H_y := H_y \cup \{(\xi_q(y,t'), t')\}$ and insert $(\xi_q(y,t'), t')$ into Q;
        Case 2. $H_y \neq \emptyset$. Let $t_1 := \max_{t'' \leq t', (\xi_q(y,t''), t'') \in H_y} t''$. If $t_1 < t'$, then let $\xi_q(y,t') := \delta(y,t')$; Otherwise, let $\xi_q(y,t') := \max k\{\xi_q(y,t_1), \delta(y,t')\}$. Insert $(\xi_q(y,t'), t')$ in H_y. If $(\xi_q(y,t'), t')$ becomes the first element in H_y, then replace $(\xi_q(y,u), u)$ of Q by this element;
      End *case*;
      Delete $(\xi_q(x,t), t)$ from Q and insert the element that is next to $(\xi_q(x,t), t)$ of H_x, if any, into Q;
  End *while*;
  Let $h = \{h_1, h_2, ..., h_k\} := \min k_{(\xi_q(\rho,t), t) \in H_\rho} \{\lceil \sigma/\xi_q(\rho,t) \rceil + t\}$;
End.

To examine H_x and delete redundant elements, we can use an fixed length queue $B_x = \{b_x^1, ..., b_x^k\}$ which contains the k maximal values of the checked $\xi_q(x,t)$ of H_x in nonincreasing order. For those $(\xi_q(x,t'), t')$ next to $(\xi_q(x,t), t)$, we compare each $\xi_q^i(x,t')$ with b_x^k $(1 \leq i \leq k)$. If $\xi_q^i(x,t') \leq b_x^k$, then delete $\xi_q^i(x,t')$ by letting $\xi_q^i(x,t') = 0$. Otherwise, insert $\xi_q^i(x,t')$ into B_x (the last one in B_x will be pushed out). If all $\xi_q^i(x,t')$ are equal to 0, then delete $(\xi_q(x,t'), t')$ in H_x and check the next element. Repeat until we find one element in H_x with at least one non-zero value, or until all elements $(\xi(x,t'), t')$ $(t' > t)$ have been deleted.

Theorem 6.6 *After the algorithm terminates, h_i $(1 \leq i \leq k)$ is the cost of the static ith quickest path from s to ρ in N.*

The proof is similar to that for Theorem 6.4, and hence we omit it here.

Theorem 6.7 *Algorithm KQP can be implemented in $O(m(r+k)(\log n + \log r + k))$ time, where r is the number of different arc capacities in N.*

Proof: The initialization step needs $O(n)$ time. Let r' be the maximum number of elements in H_x $(x \in V)$. Then, finding and deleting $\xi_q^*(x, t)$ with minimal t needs $O(1)$ time and inserting the next element of H_x (if this element exists) into Q needs $O(\log n)$ time. The step of arranging H_x requires $O(rkm)$ time. For each arc $(x, y) \in A$, calculating $\delta(y, t')$ needs $O(k)$ time. For cases 1 and 2, we need $O(\log n)$ and $O(\log r' + k)$ respectively. Since the number of elements in each H_x is less than or equal to r', the whole loop needs $O(r'm(\log n + \log r' + k))$. To compute τ we need $O(r'k)$. Thus, the total running time is bounded by $O(r'm(\log n + \log r' + k))$.

Now, we show that $r' \le (r+k)$. Note that in each H_x, all $(\xi_q(x, t), t)$ are sorted on t. By Lemma 6.4, each $\xi_q^i(x, t_j)$ must be greater than its k predecessors. Since we have r different arc capacities in total, we have $r' \le (r+k)$ elements in H_x in the worse case.

In summary, we complete the proof. □

6. Additional references and comments

Burkard, Dlaska, and Kellerer (1994) consider another quickest path problem in which the flow should be transmitted by two disjoint paths. They have shown that the problem is NP-hard in general, but is solvable in $O(m^3 n)$ time if the network is acyclic.

The time-varying maximum capacity path problem and the time-varying quickest path problem are both interesting models. One possesses the property that "any subpath of the optimal path is still optimal" while the other does not. However, as we have shown in this chapter, these two problems have a very close relationship. The time-varying quickest path can be obtained from among a set of time-varying maximum capacity paths.

Chapter 7

FINDING THE BEST PATH WITH MULTI-CRITERIA

1. Introduction

Many decision-making problems involve more than one objective to optimize. In this chapter, we will discuss some of such multicriteria optimization problems on a time-varying network. Specifically, we will focus on the situation where a path is to be sought under two objectives. The basic ideas and strategies used to tackle this type of bicriteria problems may be generalized to other cases.

The minimum cost-reliability ratio path problem, which has been studied in the literature when the network is static, is a typical bicriteria problem. This is discussed below as an example of the models we are to introduce in this chapter.

Suppose we are given a time-varying network, where a transit time $b(x, y, t)$ is needed to traverse an arc (x, y). Moreover, two attributes are associated with each arc, which are the transit cost $c(x, y, t)$ and the reliability $r(x, y, t)$, where the reliability is defined as the probability that the arc (x, y) is performable at time t. All parameters $b(x, y, t)$, $c(x, y, t)$ and $r(x, y, t)$ are the functions of the departure time t at the beginning vertex of the arc. The problem is to find a path that allows one to travel from the origin s to the destination x by a given deadline T, such that the total cost is minimum while the overall reliability of the path is maximum.

For any path $P(s = x_1, x_2, ..., x_p = x)$, its reliability is defined as:

$$\mathcal{R}(P) = \prod_{1 \leq i \leq p-1} r(x_i, x_{i+1}, \tau(x_i)).$$

Let $\delta(P) = \sum_{(x,y) \in P} g(x, y, \tau(x))$, where $g(x, y, \tau(x)) = -\ln r(x, y, \tau(x))$. Then, for any P, $\mathcal{R}(P) = \exp(-\delta(P))$. Hence, maximizing $\mathcal{R}(P)$ is equivalent to minimizing $\delta(P)$. Accordingly, the problem to minimize the total cost and maximize the overall reliability is equivalent to minimizing the following two objectives:

$$Z_1(P) = \sum_{i=1}^{p-1} c(x_i, x_{i+1}, \tau(x_i))$$

$$Z_2(P) = \sum_{i=1}^{p-1} -\ln r(x_i, x_{i+1}, \tau(x_i))$$

In the static case where it is assumed that all parameters are time independent and travel on any arc takes zero time, the problem above has been studied in the literature, which actually converts the two criteria into a single one expressed as the ratio of the cost and reliability:

$$\min_{P \in \mathcal{P}} z(P) = Z_1(P)/\mathcal{R}(\mathcal{P}).$$

Clearly, minimizing $z(P)$ achieves, to certain extent, the effect of minimizing the total cost and maximizing the reliability. Hansen (1980) presents an algorithm which can be implemented in $O(mnD \log(nD))$ time, where m and n are the numbers of arcs and vertices of the network, respectively, and $D = \max_{(x,y) \in A} \{c(x,y)\}$ ($c(x,y)$ is the cost of arc (x,y)). Ahuja (1988) presents an $O(mnD \log m)$ algorithm, under the assumption that all costs are positive.

We will consider, in this chapter, the time-varying version of the minimum cost-reliability ratio path problem. We will actually examine a bit more general model which is to minimize two criteria, both expressed as the sum of a set of cost attributes. We denote this as the MinSum-MinSum model. In addition, we will also discuss a MinSum-MinMax model, which involves another criterion expressed as the maximum in the set of cost attributes. We will lay down the concepts and assumptions on these models in Section 2, which will then be examined in the subsequent sections.

2.   Problem formulation

Consider a time-varying network, where each arc (x,y) is associated with two attributes $c^i(x,y,t)$, $i = 1,2$. The attributes may represent "cost", or other entities such as "reliability" as in the example given in Section 1 above. We assume that the transit time $b(x,y,t)$ on each arc (x,y) is a positive integer, and all $c^i(x,y,t)$ are arbitrary integers. Moreover, to simplify the illustration of key ideas in the modelling and the solution algorithms, we consider only the case where waiting at any vertex is not allowed. Consequently, we do not assume that there is any waiting cost at any vertex. Generalizations of the results in this chapter may be possible to problems involving waiting at vertex.

Let $P(s = x_1, x_2, ..., x_r = x)$ be a dynamic path from s to x in N, where s is the source vertex in N, and $Z_1(P)$ and $Z_2(P)$ are functions of P in terms of the first and the second attributes, respectively. Generally, the bicriteria problem we consider in this chapter is to determine an optimal path to minimize the two criteria $Z_1(P)$ and $Z_2(P)$:

$$\min_{P \in \mathcal{P}} Z(P) = [Z_1(P), Z_2(P)]$$

where $\mathcal{P}$ is the set of feasible paths that meet the constraint regarding the deadline T and any other constraints that may exist. We denote the problem by **TVBP** (The Time-varying bicriteria path problem). To simplify the presentation, we will assume that both $Z_1(P)$ and $Z_2(P)$ take integer values. The results we describe in this chapter may also be generalized when $Z_1(P)$ and $Z_2(P)$ are functions taking real values.

Because the problem we consider here has two objectives, finding an optimal solution of the problem, i.e., an optimal path such that both the values of $Z_1(\cdot)$ and $Z_2(\cdot)$ are minimum, may not be possible. Consequently, similar to other multicriteria optimization problems, what we should seek are the *efficient solutions* for the problem, as defined below.

Definition 7.1 *A dynamic path $\bar{P} \in \mathcal{P}$ is said to be an efficient solution for the TVBP problem if and only if there does not exist any other path $P \in \mathcal{P}$ such that $Z_1(P) \leq Z_1(\bar{P})$ and $Z_2(P) \leq Z_2(\bar{P})$ with a strict inequality in at least one case.*

Mote *et. al.* (1991) have identified two properties for the problem of finding an efficient path under two criteria, which can be extended to the time-varying problem TVBP we are considering here.

Property 7.1 *A dynamic path P from s to x is an efficient solution only if every subpath from s to an intermediate vertex of P is also an efficient solution.* □

Let $A(P)$ and $V(P)$ be the vertex set and the arc set of the path P, respectively. We will consider, more specifically, the following models in this chapter:

(1) The MinSum-MinSum problem:

$$Z_i(P) = \sum_{(x,y) \in A(P)} c^i(x, y, \tau(x)), \quad i = 1, 2.$$

(2) The MinSum-MinMax problem:

$$Z_1(P) = \sum_{(x,y)\in A(P)} c^1(x, y, \tau(x)),$$

$$Z_2(P) = \max_{(x,y)\in A(P)} \left\{ c^2(x, y, \tau(x)) \right\}.$$

Note that in some problems, it may also be desirable to have a MinMax criterion. For example, in the reliability example of Section 1, one may want to find a path where the reliability on each arc is not too low, in addition to the criterion that the overall transit cost of the path is minimum. In other words, it is desirable to minimize $\max_{(x,y)}\{-\ln r(x, y, \tau(x))\}$.

We will examine the MinSum-MinSum and MinSum-MinMax problems in Sections 3 and 4 respectively. There exists another type of problems with MinMax-MinMax criteria, which can be tackled following the similar ideas and techniques for the MinSum-MinSum and MinSum-MinMax problems.

We now introduce some operators, which we will use in the solution algorithms we will develop.

Definition 7.2 *Suppose Q_1 and Q_2 are two sets of couples, each of which has k elements sorted in nondecreasing order lexicographically. The* Merge *operator on Q_1 and Q_2 is defined as follows:*

$$Q = \text{Merge}(Q_1, Q_2)$$

where Q is a new set which contains and keeps all elements in Q_1 and Q_2 in nondecreasing order lexicographically.

For example, Let

$$Q_1 = \{(1,1), (1,2), (3,2), (4,5)\}$$

and

$$Q_2 = \{(2,2), (2,5), (4,3), (4,7)\},$$

and then

$$Q = Merge(Q_1, Q_2) = \{(1,1), (1,2), (2,2), (2,5), (3,2), (4,3), (4,5), (4,7)\}.$$

A Merge operation as defined above can be performed in $O(2k)$ time.

Definition 7.3 *Let* $Q = \{(a_1, b_1), ..., (a_l, b_l)\}$ *be a set of* l *couples, and* (a^0, b^0) *a couple.*

(i) Define an add *operator* "+" *as:*

$$Q + (a^0, b^0) = \{(a_1, b_1) + (a^0, b^0), ..., (a_l, b_l) + (a^0, b^0)\}.$$

If $Q = \emptyset$, *then define* $Q + (a^0, b^0) = \{(a^0, b^0)\}.$

(ii) Define a Max *operator* $\triangledown$ *as:*

$$Q \triangledown (a^0, b^0) = \{(\max\{a_1, a^o\}, \max\{b_1, b^o\}), ..., (\max\{a_l, a^o\}, \max\{b_l, b^o\})\}.$$

If $Q = \emptyset$, *then define* $Q \triangledown (a^o, b^o) = \{(a^o, b^o)\}.$

(iii) Define a Sum-Max *operator* $\odot$ *as:*

$$Q \odot (a^o, b^o) = \{(a_1 + a^o, \max\{b_1, b^o\}), ..., (a_l + a^o, \max\{b_l, b^o\})\}.$$

If $Q = \emptyset$, *then define* $Q \odot (a^o, b^o) = \{(a^o, b^o)\}.$

The operator "eff" is defined below, which finds all efficient points among a set of points.

Definition 7.4 *Let* $\Phi = \{(a^1, b^1), (a^2, b^2), ..., (a^k, b^k)\}$, *where* (a^i, b^i) *is a solution of the bicriteria problem TVBP,* $i = 1, 2, ..., k$. *Define* eff$\{\Phi\}$ *as an operator that identifies all efficient points of* Φ. *Define* eff$\{\Phi\} = \emptyset$ *if* $\Phi = \emptyset$.

Assume that all elements of Φ are sorted in nondecreasing order on a^i and b^i lexicographically. Then the operator "eff" can be executed by k comparisons, where k is the number of the elements in Φ.

3. The MinSum-MinSum problem

We consider, in this section, the MinSum-MinSum model. Let us introduce the following definition.

Definition 7.5 *Let* $P(s = x_1, ..., x_r = x)$ *be a dynamic path of time* t, $\zeta(x, t) = \sum_{i=1}^{r-1} c^1(x_i, x_{i+1}, \tau(x_i))$, *and* $\delta(x, t) = \sum_{i=1}^{r-1} c^2(x_i, x_{i+1}, \tau(x_i))$. *Let* $(\zeta(x, t), \delta(x, t))$ *be an efficient point of the problem, and* $\Delta_{x,t}$ *be the set of all efficient points. If there does not exist such a dynamic path* $P(s, x)$ *in* N, *let* $\Delta_{x,t} = \emptyset$.

The following lemma gives us a recursive relation to compute $\Delta_{y,t}$.

Lemma 7.1 $\Delta_{s,0} = \{(0,0)\}$ *and* $\Delta_{y,0} = \emptyset$ *for all* $y \neq s$. *For* $t > 0$ *and* $y \in V$, *we have:*

$$\Delta_{y,t} = \text{eff} \left\{ \bigcup_{\{x|(x,y)\in A\}, \{u|u+b(x,y,u)=t\}} \{\Delta_{x,u} + (c^1(x, y, u), c^2(x, y, u))\} \right\}.$$

The lemma can be proven by using the usual induction technique we have adopted in the previous chapters, based on the fact that any subpath of an efficient path will also be an efficient path (see Property 7.1).

The following algorithm can now be presented.

> **Algorithm TVBP**
> **Begin**
>   **Initialize** $\Delta_{s,0} = \{(\zeta(s,0), \delta(s,0))\} := \{(0,0)\}, \Delta_{s,t} = \emptyset$ for $0 < t \leq T$, and $\forall_{x \neq s} \Delta_{x,t} := \emptyset$ for $0 \leq t \leq T$;
>   **Sort** all values $u + b(x,y,u) = t$ for $u = 1, 2, ..., T$ and for all arcs $(x, y) \in A$;
>   **For** $t = 1, 2, ..., T$ **do**
>   **For** each $(x, y) \in A$ and each u such that $u + b(x,y,u) = t$
> **do**
> $$\bar{\Delta}^z_{y,t} := \Delta_{x,u} + (c^1(x,y,u), c^2(x,y,u));$$
> $$\Delta_{y,t} := \text{Merge}(\Delta_{y,t}, \bar{\Delta}^z_{y,t});$$
> $$\Delta_{y,t} := \text{eff}\{\Delta_{y,t}\};$$
>   **Let** $\Delta^*_z(y) = \text{eff}\{\cup_{0 \leq t \leq T} \Delta_{y,t}\}$;
> **End.**

From Lemma 7.1, we have the following result.

Lemma 7.2 *When Algorithm TVBP terminates, $\Delta^*_z(y)$ is the set of all efficient paths from s to y within the time T, and each couple $(\zeta(y,t), \delta(y,t))$ in $\Delta_{y,t}$ is an efficient path $P(s,y)$ of time t.*

Let $L = \max_{(x,y) \in A, t=0,1,...,T} c^1(x,y,t)$. The following lemma gives an estimation of the time complexity of the algorithm.

Lemma 7.3 *Algorithm TVBP can be implemented in $O(T^2 Lnm)$ time.*

Proof: The initialization needs $O(nT)$ time. To sort $u + b(x,y,u) = t$, we can use bucketsort, with T buckets. Since there are mT values to be sorted, this step needs $O(mT)$ time. During the iteration step, we need to check each arc at each time. Thus, it needs $O(mT)$ time. Noting that $|\Delta_{y,t}| \leq nTL$, computing $\bar{\Delta}^z_{y,t}$ needs at most $O(nTL)$ time. To merge $\Delta_{y,t}$ with $\bar{\Delta}^z_{y,t}$, and to obtain all efficient points of $\Delta_{y,t}$, we need at most $O(nTL)$ comparisons. Therefore, we need $O(mTnTL) = O(T^2 Lnm)$ time in the iteration step. The last step is dominated by this time requirement. In summary, the total running time is bounded by $O(T^2 Lnm)$. □

Note that we can also let $L = \max_{(x,y) \in A, t=0,1,...,T} c^2(x,y,t)$ in Lemma 7.3.

The set of efficient points is also called the efficient frontier of a multi-criteria problem. In summary, we have

Theorem 7.1 *The efficient frontier of the MinSum-MinSum problem can be obtained in* $O(T^2 Lmn)$ *time.* $\square$

4. The MinSum-MinMax problem

We now consider the MinSum-MinMax problem. Note that we assume that waiting at any vertex is not allowed.

Definition 7.6 *Let* $P(s = x_1, ..., x_r = x)$ *be a dynamic path of time* t *where waiting at any vertex* x *is prohibited, Let* $\zeta(x,t) = \sum_{i=1}^{r-1} c^1(x_i, x_{i+1}, \tau(x_i))$, *and* $\eta(x,t) = \max_{1 \le i \le r-1} c^2(x_i, x_{i+1}, \tau(x_i))$. *Let* $(\zeta(x,t), \eta(x,t))$ *be an efficient point of the problem, and* $\Delta_{x,t}^o$ *be the set of all those efficient points. If there exists no such a dynamic path* $P(s,x)$ *in* N, *let* $\Delta_{x,t}^o = \emptyset$.

The following lemma show that how to calculate $\Delta_{x,t}^o$.

Lemma 7.4 $\Delta_{s,0}^o = \{(\zeta(s,0), \eta(s,0))\} = \{(0,0)\}$ *and* $\Delta_{y,0}^o = \emptyset$ *for all* $y \ne s$. *For* $t > 0$, *we have:*

$$\Delta_{y,t}^o = \text{eff}\{ \bigcup_{\{x|(x,y)\in A\},\{u|u+b(x,y,u)=t\}} \{\Delta_{x,u}^o \odot (c^1(x,y,u), c^2(x,y,u))\}\}.$$

We leave the proof to readers. Following Lemma 7.4, we can write out the algorithm as below:

    Algorithm TVBP-MM
      Begin
        Initialize $\Delta_{s,0}^o = \{(\zeta(s,0), \eta(s,0))\} := \{(0,0)\}$, $\Delta_{s,t}^o = \emptyset$ for $0 < t \le T$, and $\forall_{x \ne s} \Delta_{x,t}^o := \emptyset$ for $0 \le t \le T$;
        Sort all values $u + b(x,y,u) = t$ for $u = 1, 2, ..., T$ and for all arcs $(x,y) \in A$;
        For $t = 1, 2, ..., T$ **do**
          For each $(x,y) \in A$ and each u such that $u + b(x,y,u) = t$
    do
        $\bar{\Delta}_{y,t}^o := \Delta_{x,u}^o \odot (c^1(x,y,u), c^2(x,y,u))$;
        $\Delta_{y,t}^o := \text{Merge}(\Delta_{y,t}^o, \bar{\Delta}_{y,t}^o)$;
        $\Delta_{y,t}^o := \text{eff}\{\Delta_{y,t}^o\}$;
      Let $\Delta_z^*(y) = \text{eff}\{\cup_{0 \le t \le T} \Delta_{y,t}^o\}$;

End.

Theorem 7.2 *The efficient frontier of the MinSum-MinMax problem can be obtained in* $O(T^2 Lmn)$ *time.*

5. Additional references and comments

A dynamic network with discrete time can be converted into a time-expanded network (TEN) without arc transit times. However, the size of this TEN is usually much greater than that of the original one. Even in the simple case where waiting at any vertex is not allowed, the numbers of vertices and arcs of such a TEN will be nT and mT, respectively. So, for Hansen's algorithm Hansen (1980), the computational time will be $O(T^2 mnL \log(nTL))$ when it is applied to the TEN, where $L = \max_{(x,y)\in A, 0\leq t\leq T} c(x,y,t)$. For Ahuja's algorithm Ahuja (1988), the time complexity will be $O(T^2 mnL \log(mT))$. As we have shown above, the time complexity of the algorithm we present is $O(T^2 Lmn)$. The reduction in the computing arises from a more direct exploitation of the problem structure.

Chapter 8

GENERALIZED FLOWS AND OTHER NETWORK PROBLEMS

1.   Introduction

In all models that we have discussed in the previous chapters, we assume that a flow travelling on an arc will retain its value all the time. In other words, the flow will not gain or lose during the transmission process. However, in some network systems, such as a water supply system, some flow may be lost during the transmission process. One can find many other examples. A flow is called a *generalized flow* if its value is increased or decreased during the transmission process. In the first part of this chapter, we will study such a time-varying network model with generalized flow.

In addition to the generalized flow problem, we will also discuss, in this chapter, other two well-known network optimization problems - the travelling salesman problem and the Chinese postman problem. The static version of the travelling salesman problem is a well-known NP-hard problem but the static Chinese postman problem is polynomially solvable. After introducing the time-varying parameters, both of the two problems will become NP-hard, as we will show in this chapter.

The organization of this chapter is as follows. We discuss the generalized flow problem in Section 2. Sections 3 and 4 consider the time-varying travelling salesman problem and Chinese postman problem, respectively. Dynamic programming approaches are described. Some additional references and remarks are provided in Section 5.

2.   Time-varying networks with generalized flows

A network problem is said to have *generalized flow* if the value of the flow travelling on an arc changes. In fact, each of the time-varying problems we have discussed in the previous chapters can be generalized by considering generalized flow. We will deal with, in this chapter, the time-varying maximum generalized flow problem. This is based on the following considerations: (1) The algorithms we will describe for this problem are quite different from those we have developed for the time-varying maximum flow problem (see Chapter 3); (2) The strategy we use to solve the problem can be adopted for solving other time-varying generalized flow problems.

2.1   Notation, assumptions, and problem formulation

The time-varying maximum generalized flow problem is to send flow from the source to the sink in a generalized network, such that the flow at the sink is maximum. A distinguished feature of such a network is that there is a positive multiplier, $\mu_{x,y}$, associated with each arc (x, y), so that if we send one unit of flow from vertex x to vertex y along the arc (x, y), then only $\mu_{x,y}$ unit arrives at vertex y. Note that in a generalized network, the flow conservation condition (Ford et al (1962)) does not hold anymore.

Consider a time-varying network $N(V, A, b, \mu, l)$. Similar to the modelling in our previous chapters, we assume that a flow must take a positive transit time $b(x, y, t)$ to traverse an arc (x, y), where t is the departure time at the beginning of the arc (x, y). All parameters, including transit times, arc capacities, vertex capacities and arc multipliers, can change over time. Given a time limit T, the problem is to find a solution to send the maximum possible flow from the source to the sink no later than T. We call this model the *time-varying maximum generalized flow* (TVMGF) problem.

Let $\mu(x, y, t)$ denote the multiplier of the arc (x, y) if the flow departs from x at time t. We assume that $\mu(x, y, t)$ can be represented by $\frac{r_1(x,y,t)}{r_2(x,y,t)}$, where $r_1(x, y, t)$ and $r_2(x, y, t)$ are two positive integers. We further assume that the capacity of arc (x, y), $l(x, y, t)$, only limits the flow which departs at vertex x at time t, which has no effect on the flow arriving at vertex y. For example, suppose $l(x, y, t) = 3$, $\mu(x, y, t) = 2$, and there is a flow waiting at x with 5 as its flow value. Thus, only 3 units of flow can be sent through the arc (x, y) at time t, since $l(x, y, t) = 3$. However, when the flow arrives at y, it will becomes 6 units, since the multiplier $\mu(x, y, t) = 2$.

Without ambiguity, we let $f(x, y, t)$ be the value of the flow departing at time t to traverse the arc (x, y), $f(x, t)$ the value of the flow waiting at vertex x during time $[t, t + 1)$ and $f(\lambda, T)$ the total flow value flowing in ρ under the solution λ, which specifies when and how to send flows from the source s to the sink ρ within the time limit T. Clearly,

$$f(\lambda, T) = \sum_{(x,\rho) \in A, t+b(x,\rho,t) \leq T} \mu(x, \rho, t) f(x, \rho, t)$$

is the amount of flow value reaching at ρ no later than the time T. The problem is to find the optimal λ^* to send a given flow v from s to ρ so as to maximize $f(\lambda^*, T)$.

As we have shown in Chapter 3, the time-varying maximum flow problem (TVMF) is NP-complete. This implies that TVMGF is also NP-complete, since TVMF is a special case of TVMGF.

In Chapter 3, we introduce the concept of dynamic f-augmenting path in a time-varying network N (Definition 3.2), which plays an important role in solving the time-varying maximum flow problem. We are now generalizing this concept to the problem TVMGF. A key issue we have to deal with is that the multipliers $\mu(x, y, t)$ should be considered. Affected by these multipliers, a flow travelling on a dynamic f-augmenting path will no longer remain as a constant. How to calculate the flow value along a dynamic f-augmenting path? Can we still calculate it by computing the capacity of the path? We will examine these questions after we introduce the definition below. To simplify, in what follows we let $l(x, t) = +\infty$ for any vertex x and time t. Extension to the case with $l(x, t) < +\infty$ can be made without much difficulty.

Definition 8.1 *Let $P(s = x_1, x_2, ..., x_r = y)$ be a dynamic f-augmenting path from s to y. Then, the capacity of path $P(s, x_i)$, while considering the multiplier factors $\mu(x, y, t)$, is recursively defined as:*

$$\begin{cases} Cap(P(s, x_i)) = \min\{Cap(P(s, x_{i-1})), l(x_{i-1}, x_i, \tau(x_{i-1}))\} \\ \qquad\qquad\qquad\qquad \cdot \mu(x_{i-1}, x_i, \tau(x_{i-1})) \qquad i = 2, 3, ..., r \\ Cap(P(s, x_1)) = +\infty. \end{cases}$$

Clearly, $Cap(P(s, y))$ gives us the upper bound of a flow which can reach each vertex x at time $\alpha(x)$ in path $P(s, y)$. Therefore, the flow value sent from s to ρ along a dynamic f-augmenting path $P(s, \rho)$ can be calculated by computing the capacity of path $Cap(P(s, \rho))$ as defined above.

Definition 8.2 *A path $P(x = x_1, ..., x_r = y)$ is called the maximum dynamic f-augmenting path from x to y of time t if $Cap(P(x, y)) \geq Cap(P'(x, y))$, where $P'(x, y)$ is any other paths from x to y of time t. Furthermore, $P(x, y)$ is said to be consistent, if for each x_i, $i = 2, ..., r$, path $P(x, x_i)$ is a maximum dynamic f-augmenting path from x to x_i.*

In the rest of this chapter, we will assume that a maximum dynamic f-augmenting path is consistent if we do not state otherwise.

2.2   Time-varying generalized residual network and properties

Recall that the basic idea of the time-varying maximum flow algorithm described in Chapter 3 is to find, repeatedly, a dynamic f-augmenting path from the source vertex to the sink vertex in the time-varying residual network and then send as much flow along the path as possible. We will use a similar strategy to solve the TVMGF problem here. Consequently, we need to generalize the concept of the time-varying residual network with the multiplier μ being taken into account.

First, for a given time-varying network N, we create a new network as follows. For every arc $(x, y) \in A$, create an artificial arc, denoted by $[y, x]$. Associated with $[y, x]$ are the transit time $b[y, x, u]$ and capacity $l[y, x, u]$. For arc $[y, x]$ and $t = 0, 1, ..., T$, let $l[y, x, t] = 0$ initially and define transit time $b[y, x, t]$ and multiplier $\mu[y, x, t]$ as follows:

$$b[y, x, t] = \begin{cases} -b(x, y, u) & 0 \leq t = u + b(x, y, u) \leq T, u = 0, 1, ..., T \\ +\infty & \text{otherwise} \end{cases}$$

$$\mu[y, x, t] = \begin{cases} 1/\mu(x, y, u) & 0 \leq t = u + b(x, y, u) \leq T, u = 0, 1, ..., T \\ 0 & \text{otherwise} \end{cases}$$

Note that $b[y, x, t]$ and $\mu[y, x, t]$ could be a multiple valued functions since for some t, there may exist more than one u satisfying $u + b(x, y, u) = t$.

For every vertex $x \in V$, we also define an artificial vertex capacity $l[x, t]$, to represent the capacity under which a flow can be "waiting" at x from time t to $t - 1$. This definition means that a flow may have a negative waiting time at a vertex x. In fact, similar to the definition of a negative transit time $b[x, y, t]$ that provides a chance to retract a flow on an arc, $l[x, t]$ provides a chance to retract a waiting time of a flow at vertex x. Initially, let $l[x, t] = 0$ for each x and $t = 1, 2, ..., T$. Obviously, the new network as created above with the initial settings is equivalent to the original one, thus we still denote it by N.

After a dynamic f-augmenting path is found, we can send an augmenting flow along it, and then construct a residual network by the following procedure:

Network Updating Procedure-UPNET

Let $N(V, A, b, \mu, l)$ be the network considered. Let $P(s, \rho) = (s = x_1, x_2, ..., x_r = \rho)$ be a dynamic f-augmenting path from s to ρ with $\tau(x_i)$, $w(x_i)$ and $\alpha(x_i)$, $f(x_i, x_{i+1}, \tau(x_i)) > 0$ the flow value sent from vertex

x_i along arc (x_i, x_{i+1}) at time $\tau(x_i)$, and $f(x_i, t)$ the flow value waiting at vertex x_i during the time $[t, t+1)$. For $i = 1, ..., r-1$, do:

Update arc capacity
   Case I: (x_i, x_{i+1}) is not an artificial arc. Let

$$l(x_i, x_{i+1}, \tau(x_i)) := l(x_i, x_{i+1}, \tau(x_i)) - f(x_i, x_{i+1}, \tau(x_i))$$

$$l[x_{i+1}, x_i, \alpha(x_{i+1})] := l[x_{i+1}, x_i, \alpha(x_{i+1})]$$
$$+ f(x_i, x_{i+1}, \tau(x_i)) \cdot \mu(x_i, x_{i+1}, \tau(x_i))$$

   Case II: (x_i, x_{i+1}) is an artificial arc. Let

$$l(x_{i+1}, x_i, \alpha(x_{i+1})) := l(x_{i+1}, x_i, \alpha(x_{i+1}))$$
$$+ f(x_i, x_{i+1}, \tau(x_i)) \cdot \mu[x_i, x_{i+1}, \tau(x_i)]$$
$$l[x_i, x_{i+1}, \tau(x_i)] := l[x_i, x_{i+1}, \tau(x_i)] - f(x_i, x_{i+1}, \tau(x_i))$$

For $i = 2, 3, ..., T-1$ do:
Update vertex capacity
   Case I: $w(x_i) > 0$. Let

$$l[x_i, t] := l[x_i, t] + f(x_i, t), \quad t = \tau(x_i), \tau(x_i) - 1, ..., \tau(x_i) - w(x_i) + 1$$

   Case II: $w(x_i) < 0$. Let

$$l[x_i, t] := l[x_i, t] - f(x_i, t), \quad t = \alpha(x_i), \alpha(x_i) - 1, ..., \alpha(x_i) + w(x_i) + 1$$

Definition 8.3 *The updated network generated by the procedure above is called a time-varying generalized residual network.*

Notice that, at each iteration of the time-varying maximum flow algorithm, the dynamic f-augmenting path found in the time-varying residual network is not necessarily a maximum one. How about the TVMGF problem now ? Let us look at the following example first.

Example 8.1

Consider a time-varying network N as shown in Figure 8.1, where the three numbers associated each arc are $b(x, y, t)$, $\mu(x, y, t)$, and $l(x, y, t)$ $(0 \le t \le 3)$. Notice that there is a dynamic f-augmenting path $P_1 = (s, e, g, \rho)$ with $Cap(P_1) = 3$. If we send the maximum possible flow along this path, we will have $f(s, e, 0) = 6$, $f(e, g, 1) = 6$, and $f(g, \rho, 2) = 3$ (which means 3 units of flow are lost during the transmission process). After that, we cannot send any more flow in N, since $f(s, e, 0) = 6 =$

Figure 8.1. Finding a dynamic f-augmenting path at each iteration may not generate the maximum generalized flow

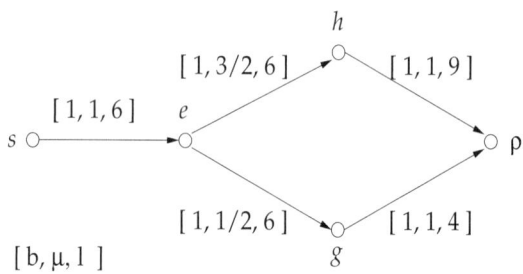

$l(s, e, 0) = 6$ and it has reached the upper bound of the arc capacity. However, one can find that another path, $P_2 = (s, e, h, \rho)$, has $Cap(P_2) = 9$. Therefore, if we send flow along path P_2, the flow value reaching ρ will be 9. Clearly, it is larger than that of P_1.

This example tells us that, only finding a dynamic f-augmenting path each time cannot allow us to obtain the maximum generalized flow, and we should find the one that has the maximum path capacity $Cap(P)$. Moreover, the problem of finding the maximum generalized flow in the original network is not equivalent to finding the maximum generalized flow in such a generalized residual network that is not created based on a maximum dynamic f-augmenting path.

We have the following property.

Property 8.1 *The problem of finding the time-varying maximum generalized flow in the original network is equivalent to finding the time-varying maximum generalized flow in the time-varying generalized residual network created based on a maximum dynamic f-augmenting path, where the equivalence is in the sense that there is a one-one correspondence between their feasible solutions.*

We omit the proof of the property here.

Definition 8.4 *Let $P(x = x_1, x_2, ..., x_r = y)$ be a dynamic path. Define*

$$\mu(P) = \prod_{1 \leq i \leq r-1} \mu(x_i, x_{i+1}, \tau(x_i))$$

as the multiplier of the path P. A path $P(x, y)$ is said to be a flow-generating path if $\mu(P) > 1$, or a flow-absorbing path if $\mu(P) < 1$. Especially, when $x_1 \neq x_2, ..., x_{r-1} \neq x_r, x_1 = x_r$ and $\tau(x_1) \geq \alpha(x_r)$, P becomes a dynamic cycle C, which is said to be a flow-generating cycle if $\mu(C) > 1$, or a flow-absorbing cycle if $\mu(C) < 1$.

Let $l_{min}(C) = \min_{(x,y) \in A(C)} l(x, y, \tau(x))$, where $A(C)$ is the arc set of C, be the minimum capacity of a dynamic cycle C. Notice that, if C is a flow-generating cycle and $l_m(C) = +\infty$, then a flow can run in C round and round while the flow value may be increased unlimitedly. This type of cycle is called *unlimited flow-generating cycle*. Clearly, if a time-varying network has such a cycle, the problem TVMGF will be unbounded. We have, however, the following property.

Property 8.2 *Neither the original network, nor the time-varying generalized residual network generated by the procedure UPNET based on a consistent maximum dynamic f-augmenting path, contains any unlimited flow-generating cycles.*

Proof: It is obvious that the original time-varying network N contains no unlimited flow-generating cycle, since all the transit times b are positive and there is no path which can visit any vertex twice with the same arrival time.

Figure 8.2. A unlimited flow-generating cycle

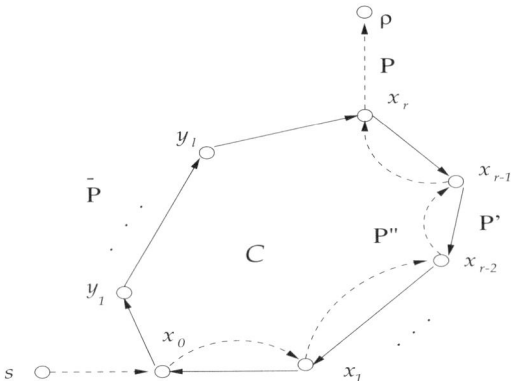

Now consider the time-varying generalized residual network. Without loss of generality, suppose that N and N' are two time-varying generalized networks, N contains no unlimited flow-generating cycles, $P(s, \rho)$ is a maximum dynamic f-augmenting path found in N, N' is created based on $P(s, \rho)$, and N' contains unlimited flow-generating cycles. Let cycle C be the one that has the minimal number of arcs among all unlimited flow-generating cycles.

See Figure 8.2, in which the dotted line represents the maximum dynamic f-augmenting path $P(s, \rho) = (s, ..., x_0, x_1, ..., x_r, ..., \rho)$ found in N. $C = (x_0, y_1, ..., y_l, x_r, x_{r-1}, ..., x_1, x_0)$ is the unlimited flow-generating cycle in N'. The section of C, $P' = (x_r, x_{r-1}, ..., x_0)$, is generated based on

$P(s, \rho)$ (see the solid line in Figure 8.2). The section $P''(x_0, x_1, ..., x_r)$ exists in N already. Let $\bar{P}$ denote the section $P(x_0, y_1, ..., x_r)$, $\mu(P'')$, $\mu(P')$ and $\mu(\bar{P})$ be the multiplier of section P'', P' and $\bar{P}$, respectively. Since C is a flow-generating cycle, we have $\mu(P') \cdot \mu(\bar{P}'') > 1$. On the other hand, since $\mu(P') = 1/\mu(P'')$, we have $\mu(P') \cdot \mu(\bar{P}) = \mu(\bar{P})/\mu(P'') > 1$, i.e., $\mu(\bar{P}) > \mu(P'')$. Therefore, we can create a new path $P^0(s, x_r)$, by combining section $(s, ..., x_0)$ of $P(s, \rho)$ and $\bar{P}$. Clearly, we have $Cap(P^0(s, x_r)) > Cap(P(s, x_r))$, where $P(s, x_r)$ is the section of path $P(s, \rho)$. Notice that $P^0(s, x_r)$ exists in N. Thus, $P(s, x_r)$ is not a maximum dynamic f-augmenting path from s to x_r in N. This contradicts the fact that $P(s, \rho)$ is a consistent maximum dynamic f-augmenting path in N. This completes the proof.   □

Similarly, we can prove the following property:

Property 8.3 *Let N' be a time-varying generalized residual network, which is created based on a maximum dynamic f-augmenting path $P(s, \rho)$, and $P'(s, \rho)$ be a maximum dynamic f-augmenting path in N'. Then, each of vertices on $P'(s, \rho)$ will not be visited twice at the same time t, where $0 \leq t \leq T$.*

2.3 Algorithms for the time-varying maximum generalized flow problem

We will now develop a pseudo-polynomial approach to solve the TVMGF problem (Note that it is NP-complete as we have indicated above). The main idea of the algorithm is to find, repeatedly, the consistent maximum dynamic f-augmenting path in the time-varying generalized residual network.

We first consider the problem under the zero waiting time constraint. A procedure, which we call the *maximum f-augmenting path searching process with the zero waiting constraint* (MDFP-ZW), will be presented below to carry out the task of searching for the maximum dynamic f-augmenting path in the time-varying generalized residual network. Similar to the procedure SDFP-ZW, it contains two different searching operations: one is a *forward searching* while the other *backward searching*. Both operations are designed by using the dynamic programming method, and forward searching is to deal with those arcs with positive transit times while backward searching will deal with negative transit times.

Definition 8.5 $\xi_z(x, t)^k$ *is the capacity of a maximum dynamic f-augmenting path from s to vertex x of time exactly t with the alter-*

nating number at most k, where the waiting time at any vertex is equal to zero.

From Property 8.3, a maximum dynamic f-augmenting path $P(s, \rho)$ cannot contain more than n vertices and each vertex cannot be visited twice or more at one time t, $t = 0, 1, ..., T$. Therefore, P cannot contain more than nT are disjoint sections. In other words, when $k \geq nT$, $\xi_z(x, t)^k$ should be the capacity of the maximum dynamic f-augmenting path from s to x of time exactly t. Recall that we denote A^+ as the set of arcs with positive transit time and A^- with negative transit time. Now, the procedure MDFP-ZW can be described as follows:

Procedure MDFP-ZW
   Initialize: $\xi_z(s, 0)^0 := +\infty, \xi_z(s, t)^0 := 0, t = 1, ..., T; \xi_z(y, t)^0 :=$
   $0, \forall y \in V \backslash \{s\}; t = 0, ..., T;$
   $i := 0;$
   Do
      $i := i + 1;$
      For all $y \in V$ and $t = 0, ..., T$ **do** $\xi_z(y, t)^i := \xi_z(y, t)^{i-1};$
      Case 1: i is an odd number:
      For $t = 1, ..., T$ **do**
         For every $y \in V$ **do** *forward searching operation:*

$$\xi_z(y, t)^i := \max\{\xi_z(y, t)^i,$$

$$\max_{\{x | (x,y) \in A^+\}} \max_{\{u | u + b(x,y,u) = t\}} \{\mu(x, y, u) \cdot \min\{\xi_z(x, u)^i, l(x, y, u)\}\}\};$$

      Case 2: i is an even number:
      For $t = T - 1, ..., 0$ **do**
         For every $y \in V$ **do** *backward searching operation:*

$$\xi_z(y, t)^i := \max\{\xi_z(y, t)^i,$$

$$\max_{\{x | (x,y) \in A^-\}} \max_{\{u | u + b[x,y,u] = t\}} \{\mu[x, y, u] \cdot \min\{\xi_z(x, u)^i, l[x, y, u]\}\}\};$$

   While there is at least one $\xi_z(y, t)^i \neq \xi_z(y, t)^{i-1};$
   Let $\xi_z^*(\rho) := \max_{0 \leq t \leq T} \xi_z(\rho, t)^i;$
   Return;

Lemma 8.1 *When the procedure MDFP-ZW terminates, $\xi_z^*(\rho)$ is the capacity of a maximum dynamic f-augmenting path from s to ρ within time T under the zero waiting time constraint.*

Proof: We only need to prove that for each i and t, $\xi_z(y,t)^i$ obtained by the procedure is the capacity of a maximum dynamic f-augmenting path from s to y of time exactly t with the alternating number at most i. Notice that any dynamic path must contain a positive section as its first section since $l[s,y,0] = 0$ for any $y \in V$ either in the original network N or in any residual networks. So, in what follows, we only consider those dynamic paths whose first sections are positive.

The proof is carried out by inductions on i. Consider $i = 1$. When $t = 0$, since $i = 1$ is an odd number, no dynamic path $P(s,y)$ of time at most 0 exists in N except when $y = s$. The capacity of the maximum dynamic path $P(s,s)$ of time 0 is ∞. In the initialization of procedure MDFP-ZW, we have $\xi_z(y,0)^1 = 0$ ($y \in V \backslash s$) and $\xi_z(s,0)^1 = \infty$; so the claim holds.

Now use the second induction on time t. When $t = 1$, there are only those paths $P(s,y)$, where vertices y is adjacent to s with $b(s,y,0) = 1$. By the definition, the capacity of $P(s,y)$ should be $l(s,y,0)\mu(x,y,0)$. By the computation of the procedure, we have

$$\xi_z(y,1)^1 = \max\{\xi_z(y,1)^1, \max_{\{y|(s,y)\in A^+\}} \{\mu(x,y,0)\cdot\min\{\xi_z(s,0)^1, l(s,y,0)\}\}\}$$

$$= \max\{0, \mu(s,y,0) \cdot \min\{\infty, l(s,y,0)\}\}$$

$$= \mu(s,y,0)l(s,y,0)$$

since $\xi_z(y,1)^1 = 0$ initially and $\xi_z(s,0)^1 = \infty$. Therefore, the claim is also true for $t = 1$.

Assume that $t > 0$ and that for all values $t' < t$ and all vertices y, $\xi_z(y,t')^1$ is the capacity of a maximum dynamic f-augmenting path $P(s,y)$ of time at most t' with the alternating number at most 1. Now, examine the case at time t.

Consider a vertex y. First we prove that, there exists a dynamic path of time at most t with the alternating number 1 and with capacity $\xi_z(y,t)^1$. By the forward searching operation, $\xi_z(y,t)^1$ must come from $\xi_z(y,t)^0$, or $\mu(x,y,u) \cdot \min\{\xi_z(x,u)^1, l(x,y,u)\}$ for some x such that $(x,y) \in A^+$ and some u such that $u + b(x,y,u) = t$. If $\xi_z(y,t)^1 = \xi_z(y,t)^0$, nothing needs to prove. If the second case occurs, by the induction on t, we know that there is a maximum dynamic f-augmenting path $P'(s = x_1, ..., x_{r-1} = x)$ from s to x of time exactly u with the alternating number at most 1 and with $Cap(P') = \xi_z(x,u)^1$. We extend the path with vertex y, obtaining a path $P(s,y)$. The time of $P(s,y)$ is exactly t, the alternating number is still 1, since the alternating number will not be changed by adding a positive arc to P', and $Cap(P(s,y)) = \mu(x,y,u) \cdot \min\{\xi_z(x,u)^1, l(x,y,u)\} = \xi_z(y,t)^1$.

We now prove that $\xi_z(y,t)^1$ is the capacity of a maximum dynamic f-augmenting path from s to y of time at most t. Let $P(s = x_1, ..., x_r = y)$ be a maximum dynamic f-augmenting path from s to y of time at most t, x the predecessor of y on this path, u the departure time of the subpath $P(s,x)$ of $P(s,y)$ at vertex x, and $Cap(P(s,x))$ the capacity of $P(s,x)$. The definition $t = u + b(x,y,u)$ implies $u < t$ since $b(x,y,u) > 0$. Thus, by induction, $Cap(P(s,x)) \geq \xi_z(x,u)^1$. By definition, the capacity of $P(s,y)$ is $Cap(P(s,y)) = \mu(x,y,u) \cdot \min\{Cap(P(s,x)), l(x,y,u)\} \geq \mu(x,y,u) \cdot \min\{\xi_z(x,u)^1, l(x,y,u)\} \geq \xi_z(y,t)^1$. We must have $Cap(P(s,y)) = \xi_z(y,t)^1$, since $P(s,y)$ is a path with maximum capacity and since there exists a path achieving $\xi_z(y,t)^1$. This completes the proof on $i = 1$.

Assume that for $i < k$, the claim is true. Now consider $i = k$.

Suppose k is an odd number. Consider the case $t = 0$ first. Since there is no dynamic path $P(s,y)$ of time exactly zero in the original or the residual networks except path $P(s,s)$, we have $Cap(P(s,y)) = 0$ and $Cap(P(s,s)) = \infty$. In procedure, we have $\xi_z(y,0)^k = \xi_z(y,0)^{k-1}$ for any $y \in V$ initially. By the induction on i, we know $\xi_z(y,0)^{k-1} = 0$ ($y \neq s$) and $\xi_z(s,0)^{k-1} = \infty$. Furthermore, these values are not be updated by the computation of the procedure. Thus, $\xi_z(y,0)^k = Cap(P(s,y))$ for any y. The claim is true.

Consider $t = 1$. If $\xi_z(y,1)^k = 0$ there is nothing to prove. Now we assume that $\xi_z(y,1)^k > 0$. For any vertex $y \in V$, by the computation of the procedure, $\xi_z(y,1)^k$ comes from $\xi_z(y,1)^{k-1}$, or $\mu(x,y,u) \cdot \min\{\xi_z(x,u)^k, l(x,y,u)\}$ for some x such that $(x,y) \in A^+$ and some u such that $u + b(x,y,u) = 1$. Suppose $\xi_z(y,1)^k$ comes from the second term and not from the first term. Since $b(x,y,u) > 0$ for any arc $(x,y) \in A^+$, we must have $u = 0$ and $b(x,y,0) = 1$. That is to say, $\xi_z(y,1)^k = \mu(x,y,0) \cdot \min\{\xi_z(x,0)^k, l(x,y,0)\}$ for some vertex x. When $x \neq s$, we know that $\xi_z(x,0)^k = 0$ (see the proof for $t = 0$). Then, $\xi_z(y,1)^k = 0$. This is contradict to the assumption of $\xi_z(y,1)^k > 0$. When $x = s$, we know $\xi_z(s,0)^k = \xi_z(s,0)^{k-1}$, therefore, $\xi_z(y,1)^k = \xi_z(y,1)^{k-1}$. This is contradict to the assumption that $\xi_z(y,1)^k$ comes from the second term only. So, we must have $\xi_z(y,1)^k = \xi_z(y,1)^{k-1}$. By the induction on i, the claim is true.

Assume the claim holds for $t' < t$. For the case where time is at t, the proof is similar to that for the case $i = 1$. One may see that clearly by replacing 1 by k. This completes the proof for the case where k is an odd number.

The proof for the case where k is an even number is similar, which we omit here.

In summary, the proof is completed. □

Lemma 8.2 *Procedure MDFP-ZW can be implemented in $O(mnT^2)$ time.*

Proof: The time needed for the initialization is bounded by $O(nT)$. The number of iterations for i is proportional to $T\sum_x\sum_{y,(y,x)\in A}1 = mT$. Since there are no flow-generating cycles in N (see Property 8.2), P cannot contain more than nT sections. In other words, $\xi_z(x,t)^k$ is the capacity of the maximum dynamic f-augmenting path from s to x of time exact t when $k \geq nT$. Therefore, the total iteration number for getting a maximum dynamic f-augmenting path is bounded by $O(mnT^2)$.   □

From Lemma 8.1 and Lemma 8.2, we have

Lemma 8.3 *When Procedure MDFP-ZW terminates, $\xi_z^*(\rho)$ obtained is the capacity of a maximum dynamic f-augmenting path from s to ρ of time at most T under the zero waiting time constraint. The procedure can be implemented in $O(mnT^2)$ time.*

Now we describe an algorithm to solve the time-varying maximum generalized flow problem with the zero waiting constraint.

> **Algorithm TVMGF-ZW**
> **Begin**
>   **Sort** all values $u + b(x,y,u)$ for $1 \leq u \leq T$ and for all arcs $(x,y) \in A^+$;
>   **Sort** all values $u + b[x,y,u]$ for $0 < u < T - 1$ and for all arcs $[x,y] \in A^-$;
> $\bar{v} := 0$;
>   **Do**
>   **Call** procedure MDFP-ZW;
>   **If** $\xi_z^*(\rho) > 0$ then let $\bar{v} := \bar{v} + \xi_z^*(\rho)$ and call the procedure UPNET; (there is a maximum dynamic f-augmenting path with flow value $\xi_z^*(\rho)$; so update the network)
>   **Else** stop; (no feasible solution to send all flow value σ from s to ρ within time T)
>   **While** $\xi_z^*(\rho) > 0$;
> **End**

The main step of the algorithm is to apply the procedure MDFP-ZW repeatedly. From Property 8.1, we have

Lemma 8.4 *When Algorithm TVMGF-ZW terminates, $\bar{v}$ is the total flow value which can reach the vertex ρ within T.*

Lemma 8.5 *The running time of Algorithm TVMGF-ZW is bounded by $O(m^2T^2\log B)$, where B is the largest integer among the multipliers*

and capacities.

Proof: We can use bucketsort for sorting, with T buckets. Since there are Tm values to be sorted, this step can be implemented in $O(Tm)$ time. Let L denote the least common denominator of the gains of a dynamic f-augmenting path. Clearly, $L \leq B^{2mT}$. Notice that, we find the maximum dynamic f-augmenting path at each iteration. Therefore, after at most mT times, the capacity of the maximum f-augmenting path in the residual network will be reduced by a factor of at least 2 (refer to Ahuja et al (1993). Thus, the total number of iterations can be bounded by $O(mT \log(B^2/B^{-2mT})) = O(mT \log B^{2mT+2}) = O(m^2T^2 \log B)$. $\square$

Combining with Lemma 8.3, we have:

Theorem 8.1 *Algorithm TVMGF-ZW solves the time-varying Maximum generalized flow problem with the zero waiting time constraint in* $O(T^4m^3n \log B)$ *time.*

The following example shows how to apply the algorithm.

Example 8.2

Consider a time-varying network as shown in Figure 8.3. The transit time $b(x, y, t)$, multiplier $\mu(x, y, t)$ and the arc capacity $l(x, y, t)$ are listed in Table 8.1. The time duration $T = 10$.

Figure 8.3. Example 8.2

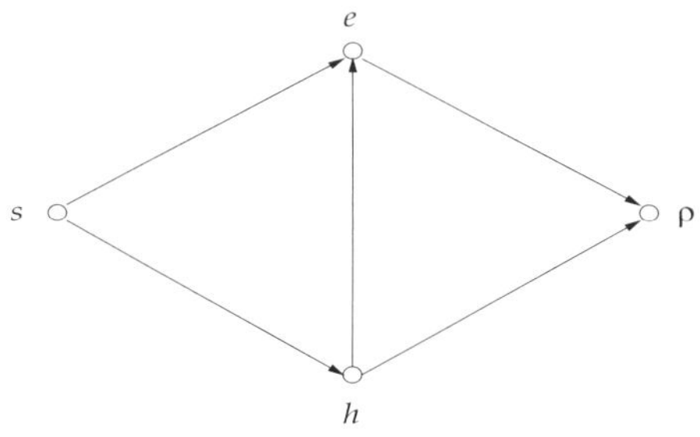

First, we call the procedure MDFP-ZW. When $i = 2$, all $\xi_z(x, t)^1 = \xi_z(x, t)^2$, then the procedure stops. We have $\xi_z(s, 0)^2 = \infty$, $\xi_z(e, 2)^2 = 12/5$, $\xi_z(e, 6)^2 = 14$, $\xi_z(h, 3)^2 = 7$, $\xi_z(\rho, 4)^2 = 72/25$, $\xi_z(\rho, 5)^2 = 15$, $\xi_z(\rho, 10)^2 = 3/7$, and all other $\xi_z(x, t)^2 = 0$. Thus, $\xi_z^*(\rho) = \max\{72/25, 15, 3/7\} = 15$. The first dynamic f-augmenting path $P_1 = (s, h, \rho)$ with $\tau(s) = 0$, $\tau(h) = 3$, and $\alpha(\rho) = 5$. The flow sent from s is 9/7, arriving at

Table 8.1. $b(x, y, t)$, $\mu(x, y, t)$ and $l(x, y, t)$

t	(s, e)	(s, h)	(e, ρ)	(h, e)	(h, ρ)
0	2, 6/5, 2	3, 7/3, 3	2, 1/2, 3	3, 6/5, 5	4, 2, 5
1	1, 1/6, 2	3, 1/2, 2	2, 4/5, 1	2, 1, 2	1, 1/3, 1
2	1, 3/4, 4	2, 7/3, 7	2, 6/5, 6	1, 3, 1	1, 5/3, 6
3	1, 5/3, 6	1, 1/6, 3	3, 6/5, 7	3, 2, 7	2, 5, 3
4	3, 3, 2	2, 2, 2	3, 1, 7	2, 1, 6	3, 4/5, 5
5	1, 1, 6	2, 3/2, 3	1, 5/2, 1	2, 1/4, 6	2, 6/5, 6
6	2, 5/7, 4	4, 6/5, 3	4, 1/7, 3	2, 4/5, 6	3, 1/2, 2
7	1, 1/2, 4	1, 5/4, 2	4, 7, 6	2, 7, 4	4, 2/3, 6
8	4, 2, 2	1, 5/6, 3	3, 2/3, 3	3, 1/2, 4	4, 4/5, 3
9	2, 1/4, 1	2, 5, 4	3, 1, 6	1, 7/4, 1	2, 1, 1
10	3, 4/3, 5	3, 1/2, 4	4, 5/4, 6	4, 1, 5	1, 1, 1

h is 3 (notice that $\mu(s, h, 0) = 7/3$), and arriving at ρ is 15 ($\mu(h, \rho, 3) = 5$).

In a similar way, we have the second path $P_2 = (s, e, \rho)$, with $\tau(s) = 0$, $\tau(e) = 2$, $\alpha(\rho) = 4$, and $\xi_z^*(\rho) = 72/25$. The third path $P_3 = (s, h, e, \rho)$, with $\tau(s) = 0$, $\tau(h) = 3$, $\tau(e) = 6$, $\alpha(\rho) = 10$, and $\xi_z^*(\rho) = 3/7$. Therefore, the total flow sent from s is $9/7 + 2 + 9/14 = 55/14$, and the total flow arriving at ρ is $15 + 72/25 + 3/7 = 3024/175$.

Now, we consider the case where $u_x = \infty$, i.e., waiting at any vertex is arbitrarily allowed.

Definition 8.6 $\xi_a(x, t)^k$ *is the capacity of a maximum dynamic f-augmenting path from s to vertex x of time at most t, with the alternating number at most k, where the waiting at any vertex has no restriction.*

The following procedure can find the maximum dynamic f-augmenting path from s to ρ of time at most t, where waiting at any vertex is arbitrarily allowed (Note that a dynamic path of time at most t is one of time at most $t + 1$ as one can wait at a vertex one unit of time more).

    Procedure MDFP-AW
      Initialize: $\xi_a(s, t)^0 := +\infty, \xi_a(y, t)^0 := 0, \forall y \in V \backslash \{s\}; t = 0, ..., T;$
      $i := 0;$
      Do
          $i := i + 1;$
           For all $y \in V, t = 0, \ldots, T$ **do** $\xi_a(y, t)^i := \xi_a(y, t)^{i-1};$
           Case 1: i is an odd number:

For $t = 1, ..., T$ **do**
  For every $y \in V$ **do** *forward searching operation:*

$$\xi_a(y, t)^i := \max\{\xi_a(y, t-1)^i, \xi_a(y, t)^i,$$

$$\max_{\{x | (x,y) \in A^+\}} \max_{\{u | u+b(x,y,u)=t\}} \{\mu(x, y, u) \cdot \min\{\xi_a(x, u)^i, l(x, y, u)\}\}\};$$

Case 2: i is an even number:
For $t = T - 1, ..., 0$ **do**
  For every $y \in V$ **do** *backward searching operation:*

$$\xi_a(y, t)^i := \max\{\xi_a(y, t+1)^i, \xi_a(y, t)^i,$$

$$\max_{\{x | (x,y) \in A^-\}} \max_{\{u | u+b[x,y,u]=t\}} \{\mu[x, y, u] \cdot \min\{\xi_a(x, u)^i, l[x, y, u]\}\}\};$$

While there is at least one $\xi_a(y, t)^i \neq \xi_a(y, t)^{i-1}$;
Let $\xi_a^*(\rho) := \xi_a(\rho, T)^i$;
Return;

Lemma 8.6 *When Procedure MDFP-AW terminates, $\xi_a^*(\rho)$ is the capacity of a maximum dynamic f-augmenting path from s to ρ of time at most T with arbitrary waiting constraint. The procedure can be implemented in $O(mnT^2)$ time.*

The proof of Lemma 8.6 is similar to that for Lemma 8.3, so we omit it here. Similar to Algorithm TVMGF-ZW, we give the algorithm below for solving the problem under the arbitrary waiting time constraints.

Algorithm TVMGF-AW
Begin
Sort all values $u + b(x, y, u)$ for $1 \leq u \leq T$ and for all arcs $(x, y) \in A^+$;
Sort all values $u + b[x, y, u]$ for $0 \leq u \leq T - 1$ and for all arcs $[x, y] \in A^-$;
  $\bar{v} := 0$;
  Do
    Call Procedure MDFP-AW;
    If $\xi_a^*(\rho) > 0$ then let $\bar{v} := \bar{v} + \xi_a^*(\rho)$ and call the procedure UPNET; (there is a maximum dynamic f-augmenting path with flow value $\xi_a^*(\rho)$; so update the network)
    Else stop; (no feasible solution to send all flow value σ from s to ρ within time T)
  While $\xi_a^*(\rho) > 0$;

End

Noting that property 8.1 and Lemma 8.5 are also true for the case of arbitrary waiting times. In summary, we have:

Theorem 8.2 *The time complexity of Algorithm TVMGF-AW is bounded above by* $O(T^4 m^3 n \log B)$.

We should point out that, when the arc multiplier $\mu(x, y, t)$ is less than or equal to 1, the time complexity of the algorithm can be improved to $O(T^3 m^3 n \log B)$. This result can be obtained by the following observation: Each maximum dynamic f-augmenting path with no artificial arcs, called the *maximum true f-augmenting path* (MTFP), cannot contain any vertex more than once. Otherwise, suppose $P = (s, ..., x_i, x_{i+1}, ..., x_i, x_j, ..., \rho)$ is a MTFP which visits vertex x_i twice. Then we can create a new path $P' = (s, ..., x_i, x_j, ..., \rho)$ by deleting the section $(x_{i+1}, ..., x_i)$ from P and waiting at vertex x_i until the departure for x_j. It is clear that P' is a feasible path with $Cap(P) \leq Cap(P')$, since P' traverses less arcs as compared to P, while $\mu \leq 1$. This contradicts the fact that P is a maximum one.

Consequently, one can see that the first k maximum f-augmenting paths obtained by the algorithm will not visit any vertex more than k times (otherwise, if one represents these k paths by k MTFPs, then there must exist a MTFP that visits the vertex more than once). Now let g_k be the number of iterations to search for the kth maximum f-augmenting path P_k, r_k the alternating number of P_k, and h_k the number of vertices of P_k. Clearly, $g_k = r_k + 1 \leq h_k$. Hence, we have $\sum_{k=1}^{v} g_k = \sum_{k=1}^{v} (r_k + 1) \leq \sum_{k=1}^{v} h_k \leq nv$. Note that $O(mT)$ is needed in each iteration. Thus, the total running time is bounded by $O(T^3 m^3 n \log B)$. In summary, we have:

Corollary 8.1 *Given a time-varying network N, if the arc multiplier $\mu(x, y, t) \leq 1$ for each arc (x, y) and each time t, the time complexity of Algorithm TVMGF-AW can be reduced to* $O(T^3 m^3 n \log B)$.

Now we examine the case with $w(x) \leq u_x, \forall x \in V$.

Definition 8.7 $\xi_b(x, t)^k$ *is the capacity of a maximum dynamic f-augmenting path from s to vertex x of time exactly t, with the alternating number at most k, where the waiting time at any intermediate vertex has an upper bound u_x.*

Similar to the two procedures MDFP-ZW and MDFP-AW, we can develop a procedure, MDFP-BW, which can find the maximum dynamic f-augmenting path in N under the bounded waiting time constraint. Notice that in the procedure presented below, the notions r_1,

$\mathcal{F}(x, y; t, r_1, r_2)$ and $J(x; u_A, u_D)$ are same as those defined in Chapter 4.

Procedure MDFP-BW
Initialize: $\xi_b(s, t)^0 := +\infty$ for $0 \leq t \leq u_s$, $\xi_b(s, t)^0 := 0$ for $u_s < t \leq T$; $\xi_b(y, t)^0 := 0, \forall y \in V\backslash\{s\}; t = 0, ..., T$; $\mathcal{R}_1(y, t) := 0, \forall y \in V; t = 0, ..., T$;
$i := 0$;
Do
$\quad i := i + 1$;
$\quad$**For** all $y \in V, t = 0, \ldots, T$ **do** $\xi_b(y, t)^i := \xi_b(y, t)^{i-1}$;
$\quad$**Case 1**: i is an odd number:
$\quad$**For** $t = 1, ..., T$ **do**
$\quad\quad$**For** every $y \in V\backslash\{s\}$ **do** *forward searching operation*:

$$\xi_b(y, t)^i := \max\{\xi_b(y, t)^i,$$

$$\max_{\{x | (x, y) \in A^+\}} \max_{(u_A, u_D) \in \mathcal{F}(x, y; t, r_1, r_2)} \{\mu(x, y, u_D) \cdot \min\{J(x; u_A, u_D), l(x, y, u_D)\}\}\};$$

$\quad$**Case 2**: i is an even number:
$\quad$**For** $t = T - 1, ..., 0$ **do**
$\quad\quad$**For** every $y \in V\backslash\{\rho\}$ **do** *backward searching operation*:

$$\xi_b(y, t)^i := \max\{\xi_b(y, t)^i,$$

$$\max_{\{x | (x, y) \in A^-\}} \max_{(u_A, u_D) \in \mathcal{F}(x, y; t, \mathcal{R}_1, r_2)} \{\mu[x, y, u_D] \cdot \min\{J(x; u_A, u_D), l[x, y, u_D]\}\}\};$$

$\quad\quad$**If** $\xi_b(y, t)^i$ is determined by the second term **then** let $\mathcal{R}_1(y, t) := r_1[x, y, u_D]$;
$\quad$**While** there is at least one $\xi_b(y, t)^i \neq \xi_b(y, t)^{i-1}$;
$\quad$**Let** $\xi_b^*(\rho) := \min_{t \in T} \xi_b(\rho, t)^i$;
$\quad$**Return**;

The main algorithm to solve the problem can now be described below:

Algorithm TVMGF-BW
Begin
$\quad$**Sort** all values $u + b(x, y, u)$ for $1 \leq u \leq T$ and for all arcs $(x, y) \in A^+$;
$\quad$**Sort** all values $u + b[x, y, u]$ for $0 \leq u \leq T - 1$ and for all arcs $[x, y] \in A^-$;
$\quad$**Calculate** $\prod_{\tau=u_A}^{u_D-1} l(x, \tau)$ and $\prod_{\tau=u_A}^{u_D+1} l[x, \tau]$ for any $x \in V$, $0 \leq u_A, u_D \leq T$;

$\bar{v} := 0;$
Do
    Call procedure MDFP-BW;
    If $\xi_b^*(\rho) > 0$ then let $\bar{v} := \bar{v} + \xi_b^*(\rho)$ and call the procedure
UPNET; (there is a maximum dynamic f-augmenting path with
flow value $\xi_b^*(\rho)$; so update the network);
    Else stop; (no feasible solution to send all flow value σ from
s to ρ within time T);
    While $\xi_b^*(\rho) > 0;$
End

Theorem 8.3 *Algorithm TVMGF-BW solves the time-varying maximum generalized flow problem under the bounded waiting time constraint in $O(T^4 m^2 n(m + nT) \log B)$ time.*

The proof is omitted here.

3.   The time-varying travelling salesman problem

The travelling salesman problem (TSP) is to determine the minimum cost path for one to visit a number of cities once and only once. This problem has many applications.

The *travelling salesman problem with time windows* (TSPTW) is an important model in time constrained routing problems, which have attracted increasing attention in recent years. In TSPTW, each vertex $x \in V$ can only be visited within a given time window $[e_x, h_x]$. The time window constraint can be divided into two types, *hard* and *soft*. In the hard case, if a vehicle arrives at a vertex too early, it will have to wait. In contrast, in the soft case, the time window constraints can be violated at a cost. The objective of the problem is to minimize the total travel cost (or time).

In this section, we consider a more general TSP: *The time-varying travelling salesman* (TVTS) problem, which can be stated as: *Given a time-varying network $N(V, A, b, c)$, where V is the vertex set, A is the arc set, $b(x, y, t)$ and $c(x, y, t)$ are the transit time and the transit cost of arc (x, y) at time t, the problem is to find a dynamic path which starts at a pre-specified vertex s, visits each vertex in V only once within time T such that the total transit cost of this path is minimized.* We assume that b is a positive integer and c is a non-negative integer.

Clearly, TSP is a special case of the TVTS problem where the transit cost c is time independent, all $b(x, y, t) = 1$, and $T = \infty$ for any arc $(x, y) \in A$ and any time $0 \le t \le T$. TSPTW is also a special case of the TVTS problem. This can be shown as follows. For the hard window constraint case, let

$$c(x, y, t) = \begin{cases} +\infty, & t < e_x \\ c(x, y), & t \geq e_x \end{cases}$$

$$c(z, x, t) = \begin{cases} +\infty, & t \geq h_x \\ c(z, x), & t < h_x \end{cases}$$

for all arcs $(z, x) \in A$ and $(x, y) \in A$, where $c(x, y)$ is the transit cost of arc (x, y) in TSPTW. For the soft window constraint case, we only need to add the extra cost to those transit costs which depart from the vertex x earlier than time e_x and arrive at x later than h_x.

It is well known that TSP is NP-hard in strong sense. Savelsbergh (1985) has shown that even finding a feasible solution of TSPTW is NP-complete. This implies that the time-varying travelling salesman problem is also NP-hard as both TSP and TSPTW are special cases of the TVTS problem.

Note that there is a dynamic programming approach (see Ahuja et al (1993)) which can optimally solve the static TSP in $O(n^2 2^n)$ time, where n is the number of vertices. If we apply it to the time-expanded network (a static network converted from the multi-period dynamic network) for solving the TVTS problem, the running time will be $O((nT)^2 2^{nT})$, since there are $O(nT)$ vertices in the time-expanded network corresponding to the time-varying network we consider. In what follows, we will describe a dynamic programming approach directly on the original time-varying network, which can be implemented in $O(mnT2^n)$ time. Clearly, this is much better than $O((nT)^2 2^{nT})$, especially when T is a large number.

We first examine the case with no constraint on waiting at any vertex, and then develop algorithms for the TVTS problem with zero waiting and bounded waiting constraints, respectively.

Definition 8.8 *Given a vertex set $S \subset V$ and $s, x \in S$, define $D_a(x, S, t)$ as the cost of the shortest dynamic path in S which starts at vertex s, visits all other vertices of S exactly once, and ends at vertex x. Waiting at vertex is allowed, but the time of this path will not excess t. If there is no such a dynamic path in S, define $D_a(x, S, t) = \infty$.*

Lemma 8.7 *For any time t we have $D_a(s, S = \{s\}, t) = 0$. For any vertex $y \neq s$, if $s \notin S$, we have $D_a(y, S, t) = \infty$ for any time t. Otherwise, if $s \in S$, we have $D_a(y, S, 0) = \infty$ and*

$$D_a(y, S, t) = \min\{D_a(y, S, t - 1),$$

$$\min_{x \in S \setminus \{y, s\}} \min_{(x, y) \in A} \min_{\{u | u + b(x, y, u) = t\}} \{D_a(x, S \setminus \{y\}, u) + c(x, y, u)\}\}$$

for any $0 < t \leq T$.

Proof: By the definition, $D_a(s, S, t) = 0$ is true when $S = \{s\}$ for any time t. Furthermore, for any $y \neq s$ and $s \notin S$, we have $D_a(y, S, t) = \infty$, since the dynamic path in S, if any, does not start at s.

Now, we consider the case $y \neq s$ and $s \in S$. When time $t = 0$, $D_a(y, S, 0) = \infty$ is obviously, since the transit time is positive and there is no path from s to y within time 0. When $t = 1$, if there is a dynamic path from s to y within time 1, then we must have $S = \{s, y\}$ and $b(s, y, 0) = 1$, because all transit times are positive integers. From the formula, $D_a(y, S, 1) = c(s, y, 0)$. Otherwise, if there is no dynamic path from s to y within time 1 (actually, there is no arc (s, y) in A such that $b(s, y, 0) = 1$), $D_a(y, S, 1) = D_a(y, S, 0) = \infty$. Therefore, the claim holds.

Assume that for any time $t' < t$, the claim is true. Consider the case of time t.

If $D_a(y, S, t) = \infty$, then by the induction, we know that there is no dynamic path from s to $x \in S \setminus \{y\}$ in S within time $u = t - b(x, y, u)$, therefore, no dynamic path from s to y within time t. Now, we assume that $D_a(y, S, t) < \infty$. First, we show that there is a dynamic path from s to y in S within time t. From the formulation, if $D_a(y, S, t) = D_a(y, S, t - 1)$, then by the induction, there is a dynamic path from s to y in S within time $t - 1$. Clearly, it is also a dynamic path within time t. Otherwise, if $D_a(y, S, t) = D_a(x, S \setminus \{y\}, u) + c(x, y, u)$, then we have $D_a(x, S \setminus \{y\}, u) < \infty$. It means that there is a dynamic path $P(s, x)$ in S within time u. Therefore, we can add arc (x, y) to P and obtain a dynamic path $P(s, x, y)$ in S within time t.

Now, we prove that $D_a(y, S, t)$ is the cost of the shortest dynamic path from s to y. Suppose that P^* is the optimal path from s to y in S within time t and $\zeta(P^*)$ is its cost. Clearly, we have $\zeta(P^*) \leq D_a(y, S, t)$. On the other hand, assume that $P^*(s, x)$ is the subpath of P^* in $S \setminus \{y\}$ from s to x with arrival time u^* at x. Therefore, we have $\zeta(P^*(s, x)) \geq D_a(x, S \setminus \{y\}, u^*)$ by the induction. That is to say $\zeta(P^*) = \zeta(P^*(s, x)) + c(x, y, u^*) \geq D_a(x, S \setminus \{y\}, u^*) + c(x, y, u^*) \geq D_a(y, S, t)$. In summary, we have $\zeta(P^*) = D_a(y, S, t)$. This completes the proof.   □

Now we are ready to describe the algorithm.

    Algorithm TVTSP-AW
    Begin
      Sort all values $u + b(x, y, u) = t$ for $u = 1, 2, ..., T$, and for all arcs $(x, y) \in A$;
      For $t = 0, 1, ..., T$ **do** $D_a(s, \{s\}, t) := 0$;

For $l = 1, 2, ..., n-1$ **do**
  For each subset $\bar{S} \subset V\backslash\{s\}$ such that $|\bar{S}| = l$ **do**
   $S := \bar{S} + \{s\}$;
    For $t = 1, 2, ..., T$ **do**
      For any $y \in \bar{S}$ **do**
       $D_a(y, S, 0) := \infty$;
        For each vertex $x \in \bar{S}\backslash\{y\}$, $(x, y) \in A$ and each u such that $u + b(x, y, u) = t$ **do**

$$D_a(y, S, t) := \min\{D_a(y, S, t-1),$$

$$\min_{x\in\bar{S}\backslash\{y\}} \min_{(x,y)\in A} \min_{\{u|u+b(x,y,u)=t\}} \{D_a(x, S\backslash\{y\}, u) + c(x, y, u)\}\};$$

  Let $z^* := \min_{y\in V\backslash\{s\}} D_a(y, V, T)$;
  End.

Theorem 8.5 *After the termination of Algorithm TVTSP-AW, z^* is the cost of the shortest dynamic path which starts at s and visits each vertex of V once and only once within the time limit T.*

Theorem 8.5 follows from Lemma 8.7 directly.

Lemma 8.8 *Algorithm TVTSP-AW can be implemented in $O(mnT2^n)$ time.*

Proof: At the sorting step we need $O(mT)$ time and the initial setting for $D_a(s, S, t)$ needs $O(T)$ time. To calculate each subset $\bar{S} \subset V\backslash\{s\}$ we need $C_{n-1}^1 + C_{n-1}^2 + ... + C_{n-1}^{n-1} \leq 2^n$ time, i.e., $O(2^n)$ time. The sub-loop of calculating $D_a(y, S, t)$ needs $O(mnT)$ time. So, the whole loop needs $O(mnT2^n)$ time. The last step is to calculate z^* which needs $O(n)$ time. In summary, we need $O(mnT2^n)$ time in total. $\qquad\square$

Based on Lemmas 8.7 and 8.8, we have

Theorem 8.6 *Algorithm TVTSP-AW can optimally solve, in $O(mT2^n)$ time, the TVTS problem with no waiting time constraint at vertices.*

Similarly, we can develop algorithms for the TVTS problem under zero waiting time and bounded waiting time constraints, respectively.

Definition 8.9 *Given a vertex set $S \subset V$ and $s \in S$, define $D_z(x, S, t)$ as the cost of the shortest dynamic path in S which starts at vertex s, visits all other vertices of S exactly once, ends at vertex x. Waiting at vertex is strictly prohibited, and the time of this path is t. If there is no such a dynamic path in S, define $D_z(x, S, t) = \infty$.*

Lemma 8.9 $D_z(s, S = \{s\}, 0) = 0$, *and For any time* $t > 0$, *we have* $D_z(s, S = \{s\}, t) = \infty$. *For any vertex* $y \neq s$, *if* $s \notin S$, *we have* $D_z(y, S, t) = \infty$ *for any time* t. *Otherwise, if* $s \in S$, *we have* $D_z(y, S, 0) = \infty$ *and*

$$D_z(y, S, t) = \min_{x \in S \setminus \{y, s\}} \min_{(x,y) \in A} \min_{\{u | u + b(x,y,u) = t\}} \{D_z(x, S \setminus \{y\}, u) + c(x, y, u)\}.$$

The following is the algorithm to solve the problem with zero waiting time constraint.

> **Algorithm TVTSP-ZW**
> **Begin**
> **Initialize**: $D_z(s, \{s\}, 0) = 0$, $D_z(s, \{s\}, t) := \infty$ for $t = 1, ..., T$;
> **Sort** all values $u + b(x, y, u) = t$ for $u = 1, 2, ..., T$, and for all arcs $(x, y) \in A$;
> **For** $l = 1, 2, ..., n - 1$ **do**
> > **For** each subset $\bar{S} \subset V \setminus \{s\}$ such that $|\bar{S}| = l$ **do**
> > $S := \bar{S} + \{s\}$;
> > **For** $t = 1, 2, ..., T$ **do**
> > > **For** any $y \in \bar{S}$ **do**
> > > $D_z(y, S, t) := \infty$;
> > > > **For** each $x \in \bar{S} \setminus \{y\}$, $(x, y) \in A$ and each u such that $u + b(x, y, u) = t$ **do**

$$D_z(y, S, t) := \min_{x \in \bar{S} \setminus \{y\}} \min_{(x,y) \in A} \min_{\{u | u + b(x,y,u) = t\}} \{D_z(x, S \setminus \{y\}, u) + c(x, y, u)\};$$

> **Let** $z^* := \min_{y \in V \setminus \{s\}} \min_t D_z(y, V, t)$;
> **End.**

Theorem 8.7 *Algorithm TVTSP-ZW can optimally solve the TVTS problem with the zero waiting time constraint in* $O(mnT2^n)$ *time.*

Definition 8.10 *Given a vertex set* $S \subset V$ *and* $s \in S$, *define* $D_b(x, S, t)$ *as the cost of the shortest dynamic path in* S *which starts at vertex* s, *visits all other vertices of* S *exactly once, and ends at vertex* x. *Waiting at vertex* x *is allowed, however, it is bounded by a given number* u_x, *and the time of this path must be* t. *If there is no such a dynamic path in* S, *define* $D_b(y, S, t) = \infty$.

Lemma 8.10 $D_b(s, S = \{s\}, t) = 0$ *for* $0 \leq t \leq u_s$ *and* $D_b(s, S = \{s\}, t) = \infty$ *for* $u_s < t \leq T$. *For any vertex* $y \neq s$, *if* $s \notin S$, *we have* $D_b(y, S, t) = \infty$ *for any time* t. *Otherwise, if* $s \in S$, *we have* $D_b(y, S, 0) = \infty$ *and*

$$D_b(y, S, t) =$$

$$\min_{x \in S \setminus \{y,s\}} \min_{(x,y) \in A} \min_{(u_A,u_D) \in \mathcal{F}(x,y,t)} \{D_b(x, S \setminus \{y\}, u_A) + c(x,y,u_D)\}$$

where $\mathcal{F}(x,y,t) = \{(u_A, u_D) | u_D + b(x,y,u_D) = t, 0 \le u_D - u_A \le u_x\}$.

Algorithm TVTSP-BW
Begin
  Initialize: $D_b(s, S = \{s\}, t) = 0$ for $0 \le t \le u_s$, $D_b(s, S = \{s\}, t) = \infty$ for $u_s < t \le T$;
  Sort all values $u + b(x,y,u) = t$ for $u = 1, 2, ..., T$, and for all arcs $(x,y) \in A$;
  For $l = 1, 2, ..., n-1$ **do**
    For each subset $\bar{S} \subset V \setminus \{s\}$ such that $|\bar{S}| = l$ **do**
    $S := \bar{S} + \{s\}$;
      For $t = 1, 2, ..., T$
  do
        For each $y \in \bar{S}$ **do**
        $D_b(y, S, t) := \infty$;
          For each vertex $x \in \bar{S} \setminus \{y\}$, $(x,y) \in A$ and each (u_A, u_D)
  such that $(u_A, u_D) \in \mathcal{F}(x,y,t)$ **do**
          $D_b(y, S, t) :=$

$$\min_{x \in \bar{S} \setminus \{y\}} \min_{(x,y) \in A} \min_{(u_A,u_D) \in \mathcal{F}(x,y,t)} \{D_b(x, S \setminus \{y\}, u_A) + c(x,y,u_D)\};$$

  Let $z^* := \min_{y \in V \setminus \{s\}} \min_t D_b(y, V, t)$;
  End.

Theorem 8.8 *Algorithm TVTSP-BW can optimally solve the TVTS problem with the bounded waiting time constraint in* $O(mnT2^n \log T)$ *time.*

4. The time-varying Chinese postman problem

The *Chinese Postman Problem* (CCP) is another important network optimization problem. CPP aims to find a tour which visits each arc of a given network at least once such that the total cost is minimized.

We will investigate, in this section, the time-varying Chinese postman (TVCP) problem, which can be stated as: *Given a time-varying network* $N(V, A, b, c)$, *where* V *is the vertex set,* A *is the arc set,* $b(x,y,t)$ *and* $c(x,y,t)$ *are the transit time and the transit cost of arc* (x,y) *at time* t, *respectively, find a dynamic path starting from a specified vertex* s *and visiting each arc at least once within a given time duration* T, *such that the total transit cost is minimized.*

Definition 8.12 *A dynamic path that visits each arc of the given network at least once is called a dynamic Chinese path. If the end vertex* *of the path is the same as the start vertex of the path, then this path is called dynamic Chinese tour.*

4.1   NP-hardness analysis

It is known that the static CPP is polynomially solvable. The TVCP problem is, however, NP-hard in the strong sense.

Theorem 8.12 *The TVCP problem is NP-hard in the strong sense.*

Proof: We transform the directed Hamilton path problem to TVCP.

Directed Hamilton Path (DHP): *Given a digraph $G = (V, A)$, does G contain a directed Hamilton path?*

Decision Version of TVCP: *Given a network $N(V, A, b, c)$ and a number K, does N contain a dynamic Chinese path such that the total cost is less than or equal to K?*

For a given digraph $G = (V, A)$, we create a network $N(V', A', b, c)$ as follows: Split each vertex $x \in V$ into two vertices x' and x'', and create an arc (x', x'') to connect them. For each arc $(y, x) \in A$ create an arc (y'', x'), and for each arc $(x, z) \in A$, create an arc (x'', z'). Moreover, create a super vertex, v, and for each arc $(x, y) \in A'$, create two arcs (v, x) and (y, v). Lastly, for each vertex $x \in A$ with $d^-(x) = 0$, create an arc (x'', s). Combining all x', x'' and v to form V' and all arcs we obtain A'. Let $b(x, y, t) = 1$ for each arc $(x, y) \in A'$ and for each time t. Let

$$c(x, y, t) = 0, \forall (x, y) \in A', (x, y) \neq (x', x''), 0 \leq t \leq T$$

$$c(x', x'', t) = 0, 0 \leq t \leq 2n - 1$$

$$c(x', x'', t) = 2, 2(n - 1) < t \leq T$$

$K = 1$ and $T = 3m + 2n + 1$. See Figure 8.4.

Obviously, the construction step can be completed in polynomial time. Now, we show that a "yes" answer for DHP is equivalent to a "yes" answer for the TVCP problem.

If G has a directed Hamilton path $P(x_1, x_2, ..., x_n)$, then we also have a dynamic path $P'(x'_1, x''_1, x'_2, x''_2, x'_3, ..., x''_n)$ in N. It means that starting from x'_1, we can visit all those arcs (x'_i, x''_i) $(i = 1, ..., n)$ before time $t = 2n - 1$, since all arc transit times are equal to one. The cost of path P' is zero, since $c(x'_i, x''_i, t) = 0$ when $t \leq 2n - 1$. Consider x_n. If $d^-(x_n) = 0$ in G, we can extend path P' to P'' by adding an arc (x''_n, s) to P', where $d^-(x)$ is the in-degree of vertex x in N. Otherwise, if $d^-(x_n) > 0$ then there must be an arc (x_n, x_j) in A. Therefore, we can extend P' to P''

Figure 8.4. Construct a time-varying network N for a given digraph G

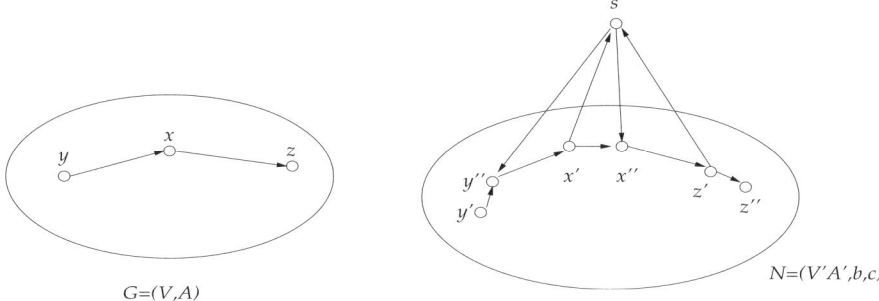

by adding arcs (x_n'', x_j') and (x_j', s) to P'. Notice that all other unvisited arcs in A' can be visited by a path which starts from vertex s. Denote this path as $\bar{P}$. Combining P'' and $\bar{P}$ we obtain a dynamic Chinese path of N. Since $\zeta(\bar{P}) = 0$ and $\zeta(P'') = 0$, the cost of the dynamic Chinese path is also zero. Furthermore, since $\tau(P'') \leq 2n+1$ and $\tau(\bar{P}) \leq 3m$, the time the dynamic Chinese path is less than or equal to $3m + 2n + 1 = T$.

If N has a dynamic Chinese path which can visit all arcs in A' within time T with cost $K' \leq K = 1$, all those arcs $(x_i', x_i'') \in A'$ must be visited before time $t = 2n - 1$, since the cost of those arcs should be 2 after that time. This condition guarantees that there a dynamic path $P'(x_1', x_1'', x_2', x_2'', x_3', ..., x_n'')$ in N. Therefore, we can find a corresponding directed Hamilton path in A.

In summary, we complete the proof. □

4.2 Dynamic programming

We now develop an exact algorithm to solve the TVCP problem. We first study the problem with no waiting constraint at any vertex, and then describe an algorithm for solving the problems with waiting time constraints at vertices.

To simplify the formulation, we introduce an extra vertex s_0 and an extra arc $r = (s_0, s)$, and let both its cost and transit time be equal to zero. r can be regarded as a *source arc* in the given network N. A dynamic path is said to travel from arc $a_1 = (x_1, y_1)$ to arc $a_2 = (x_2, y_2)$, denoted by $P(a_1, a_2)$, if a_1 is its starting arc and a_2 the ending arc. Moreover, we denote $\tau(a_1) = \tau(x_1)$ as the departure time of arc a_1, $\alpha(a_1) = \alpha(y_1)$ as the arrival time of arc a_1, $b(a_1, t) = b(x_1, y_1, t)$ and $c(a_1, t) = c(x_1, y_1, t)$. In this section, the notions $P(a_1, a_2)$ and $P(x_1, y_2)$ are used interchangeably.

Definition 8.13 *Given an arc set $M \subset A$ and $r \in M$, define $Q_a(e, M, t)$ as the cost of a shortest dynamic Chinese path from r to e of time at most t in M, where waiting at a vertex is arbitrarily allowed. If such a dynamic Chinese path dose not exist in M, let $Q_a(e, M, t) = \infty$.*

Let $z = \min_{e \in A} Q_a(e, A, T)$, which is the cost of the shortest dynamic Chinese path of time at most T. Given e, M, and t, a dynamic Chinese path from r to e within time t can be obtained by merging a dynamic Chinese path from arc r to arc $g \in M \backslash \{e\}$ within time $u < t$ and a shortest dynamic path from g to e which $\tau(g) \geq u$ and $\alpha(e) \leq t$. We have the following recursive relation.

Lemma 8.11 *For any time $0 \leq t \leq T$, $Q_a(r, M = \{r\}, t) = 0$. For any arc $e \in M$ and $e \neq r$, if $r \notin M$, $Q_a(e, M, t) = \infty$ for any time t; otherwise, if $r \in M$, $Q_a(e, M, 0) = \infty$ and*

$$Q_a(e, M, t) = \min\{Q_a(e, M, t-1),$$

$$\min_{g \in M \backslash \{e\}} \min_{\{u | \tau(g) \geq u, \alpha(e) \leq t, 0 \leq u < t\}} \{Q_a(g, M \backslash \{e\}, u) + \zeta(P_M(g, e))\}\}$$

for $0 < t \leq T$, where $P_M(g, e)$ is the shortest dynamic path from g to e in M, $\zeta(P_M(g, e))$ is its cost, $\tau(g)$ and $\alpha(e)$ are the departure time at arc g and the arrival time at arc e in path $P_M(g, e)$, respectively.

Proof: When $M = \{r\}$, it is obvious that $Q_a(r, M, t) = 0$ for any t. For any arc $e \in M$ and $e \neq r$, if $r \notin M$, $Q_a(e, M, t) = \infty$ holds since there is no dynamic path in M which starts from arc r. In what follows, we consider the case $r \in M$.

When time $t = 0$, $Q_a(e, M, 0) = \infty$ is true, since all the transit times are positive numbers and there is no dynamic path including arcs r and e within time 0. For any time $t > 0$ and $t \leq T$, we prove the recursive relation by induction.

When $t = 1$, if there is a dynamic Chinese path in M of time at most 1, then must have $M = \{r, e\}$ and $u = 0$ and $b(e, 0) = 1$, since all the transit times are positive integers. Therefore, the cost of the shortest dynamic Chinese path from r to e within time 1 is $c(e, 0)$. On the other hand, by the recursive relation, we have $Q_a(e, M, 1) = \min\{Q_a(e, M, 0), Q_a(g, M \backslash \{e\}, 0) + \zeta(P_M(g, e))\} = c(e, 0)$. Thus the claim is true.

Assume that the claim is true for any time less than t. Now consider the case of time t.

If $Q_a(e, M, t) = \infty$, then there is nothing to prove. Now assume $Q_a(e, M, t) < \infty$. First, we prove that there is a dynamic Chinese path of time at most t in M which starts at arc r and ends at arc e. If

$Q_a(e, M, t)$ comes from $Q_a(e, M, t-1)$, by the induction on t, we know that there is a dynamic Chinese path of time at most $t-1$, of course, it is a path of time at most t. Otherwise, if $Q_a(e, M, t)$ comes from the second term only, that is to say, there is a dynamic Chinese path of time at most u in $M \backslash \{e\}$, denoted by $P(r, g)$, with g as its ending arc. Moreover, there is a dynamic shortest path $P_M(g, e)$ with $\tau(g) \geq u$ and $\alpha(e) \leq t$. Now we can combine paths $P(r, g)$ and $P(g, e)$ to obtain a new path $P(r, e)$. Obviously, it is a dynamic Chinese path of time at most t in M.

We now prove that $Q_a(e, M, t)$ is the value of the shortest dynamic Chinese path in M. Suppose that $P^*(r, e)$ is the optimal path and $\zeta(P^*)$ is its cost. Then we have $\zeta(P^*) \leq Q_a(e, M, t)$. On the other hand, let h be such an predecessor arc of e on path P^* that (i) all arcs in $M \backslash \{e\}$, except h, have been visited before h; (ii) h is first time visited. Let $P^*(r, h)$ and $P^*(h, e)$ be the subpaths of P^* from r to g with time at most u' and from h to e, respectively. Therefore, by the induction, we have $\zeta(P^*(r, h)) \geq Q_a(h, M \backslash \{e\}, u')$, since $u' < t$. Moreover, we have $\zeta(P_M(h, e) \leq \zeta(P^*(h, e))$, since $P_M(h, e)$ is the shortest dynamic path in M. That means $\zeta(P^*(r, e)) \geq Q_a(h, M \backslash \{e\}, u') + \zeta(P_M(h, e)) \geq Q_a(e, M, t)$. The last inequality comes from the formula. In summary, we have $\zeta(P^*(r, e)) = Q_a(e, M, t)$. This completes the proof. □

Definition 8.14 *Let $M \subset A$. Define $R(M, x, t)$ as the cost of a shortest dynamic Chinese path in M that starts from s and ends at vertex x within time t. If such a path does not exist, let $R(M, x, t) = \infty$.*

For a given arc set $M \subset A$, let $N[M]$ denote the arc induced subnetwork of N which is generated by M on N, and $V(N[M])$ be the vertex set of $N[M]$. We have:

Lemma 8.12 *For each $M \subset A$, each vertex $x \in V(N[M])$ and each time $0 \leq t \leq T$, we have*

$$R(M, x, t) =$$

$$\min_{e=(y,z)\in M} \min_{\{u|\tau(z)\geq u, \alpha(x)\leq t, 0\leq u\leq t\}} \{Q_a(e, M, u) + \zeta(P_{N[M]}(z, x))\},$$

where $P_{N[M]}(z, x)$ is the dynamic shortest path in subnetwork $N[M]$.

Lemma 8.12 follows from Definition 8.14 directly. By Lemma 8.12, the calculation shown in Lemma 8.11 can be rewritten as

$$Q_a(e, M, t) :=$$

$$\min\{Q_a(e, M, t-1), \min_{\{u|u+b(x,y,u)=t\}} \{R(M \backslash \{e\}, x, u) + c(x, y, u)\}\},$$

where $e = (x, y) \in M$.

To obtain $R(M, x, t)$, we can use a method as follows: Suppose $e = (y, z)$. To obtain $Q_a(e, M, t)$, we can calculate all shortest paths from z to each other vertex $x \in V(N[M])$ in $N[M]$, with the departure time t and the arrival time t', where $t \leq t' \leq T$. Then, we update the value of $R(M, x, t')$ ($t \leq t' \leq T$) by letting $R(M, x, t') = \min\{R(M, x, t'), Q_a(e, M, t) + \zeta(P_{N[M]}(z, x))\}$, where $\zeta(P_{N[M]}(z, x))$ is the value of the shortest path of $P_{N[M]}(z, x)$. Initially, let $R(M, x, t') = \infty$ for any time $0 \leq t' \leq T$.

We now have the following algorithm.

> **Algorithm TVCPP-AW**
> **Begin**
>   **Sort** all values $u + b(x, y, u) = t$ for $u = 1, 2, ..., T$, and for all arcs $(x, y) \in A$;
>   **For** $t = 0, 1, ..., T$ **do** $Q_a(r, \{r\}, t) := 0$, $Q_a(e, \{r\}, t) := \infty$;
>   **For** $l = 1, 2, ..., n - 1$ **do**
>   **For** each subset $\bar{M} \subset A \backslash \{r\}$ such that $|\bar{M}| = l$ **do**
> $M := \bar{M} + \{r\}$;
>   **For** each $x \in V(N[M])$ and $t = 0, 1, ..., T$ **do** $R(M, x, t) :=$ ∞;
>   **For** each $e \in \bar{M}$ **do** $Q_a(e, M, 0) := \infty$;
>   **For** $t = 1, 2, ..., T$ **do**
>   **For** each $e = (x, y) \in \bar{M}$ **do**
> $Q_a(e, M, t) :=$
>
> $$\min\{Q_a(e, M, t - 1), \min_{\{u | u + b(x, y, u) = t\}} \{R(M \backslash \{e\}, x, u) + c(x, y, u)\}\};$$
>
>   Call procedure TSP-AW to calculate all paths from y to each other vertex $z \in V(N[M])$ and update $R(M, z, t')$;
>   **Let** $z^* := \min_{e \in A} Q_a(e, A, T)$;
> **End.**

Lemma 8.13 *After the termination of Algorithm TVCPP-AW, z^* is the value of the shortest dynamic Chinese path of N within time T.*

Lemma 8.13 follows from Lemma 8.11 and 8.12 directly.

Lemma 8.14 *Algorithm TVCPP-AW can be implemented in $O(m^2 n T^2 2^m)$ time.*

Proof: The sorting step needs $O(mT)$ time. Setting initial values for $Q_a(e, \{r\}, t)$ needs $O(T)$ time. Examining each subset $\bar{M}$ of $A \backslash \{r\}$ needs $O(2^m)$ time. For each arc set $M = \bar{M} + \{r\}$, initializing $R(M, x, t)$ for each vertex $x \in V(N[M])$ needs $O(nT)$ time. For each arc $e \in \{M\}$ to set $Q_a(e, M, 0)$ needs $O(m)$ time. Calculating $Q_a(e, M, t)$ needs $O(mT)$ time, since we have sorted all u already. Computing all shortest paths from y (the end of arc e) to each vertex $z \in V(N[M])$ and updating $R(M, z, t')$ need $O(mnT^2(m+n))$ and $O(mnT^2)$ respectively. Therefore the total running time of the algorithm is bounded by $O(m^2nT^22^m)$. □

Theorem 8.13 *The TVCP problem with the arbitrary waiting constraint can be solved in $O(m^2nT^22^m)$ time.*

Now, we consider the problem with zero waiting and bounded waiting constraints.

Definition 8.15 *Given an arc set $M \subset A$ and $r \in M$, define $Q_z(e, M, t)$ as the cost of a shortest dynamic Chinese path from r to e of time exactly t in M, where waiting at a vertex is strictly prohibited. If such a dynamic Chinese path dose not exist in M, let $Q_z(e, M, t) = \infty$.*

Lemma 8.15 $Q_z(r, M = \{r\}, 0) = 0$ *and for any time $0 < t \leq T$, $Q_z(r, M = \{r\}, t) = \infty$. For any arc $e \in M$ and $e \neq r$, if $r \notin M$, $Q_z(e, M, t) = \infty$ for any time t; otherwise, if $r \in M$, $Q_z(e, M, 0) = \infty$ and*

$$Q_z(e, M, t) =$$
$$\min_{g \in M \setminus \{e\}} \min_{\{u \mid \tau(g) = u, \alpha(e) = t, 0 \leq u \leq t\}} \{Q_z(g, M \setminus \{e\}, u) + \zeta(P_M(g, e))\}\}$$

for $0 < t \leq T$.

Definition 8.16 *Let $M \subset A$. Define $R_z(M, x, t)$ as the cost of a shortest dynamic Chinese path in M that ends at vertex x at time t, where waiting at any vertex is strictly prohibited. If this path does not exist, let $R_z(M, x, t) = \infty$.*

Lemma 8.16 *For each $M \subset A$, each vertex $x \in V(N[M])$ and each time $0 \leq t \leq T$, we have*

$$R_z(M, x, t) =$$
$$\min_{e=(y,z) \in M} \min_{\{u \mid \tau(z) = u, \alpha(x) = t, 0 \leq u \leq t\}} \{Q_z(e, M, u) + \zeta(P_{N[M]}(z, x))\}.$$

The calculation shown in Lemma 8.15 can be rewritten to

$$Q_z(e, M, t) := \min_{\{u \mid u + b(x,y,u) = t\}} \{R_z(M \setminus \{e\}, x, u) + c(x, y, u)\},$$

where $e = (x, y) \in M$.

Algorithm TVCPP-ZW
Begin

Initialize: $Q_z(r, \{r\}, 0) := 0$ and $Q_z(r, \{r\}, t) := \infty$ for $0 < t \leq T$,
$Q_z(e, \{r\}, t) := \infty$ for any $e \neq r$ and any time $0 \leq t \leq T$;
    Sort all values $u + b(x, y, u) = t$ for $u = 1, 2, ..., T$, and for all arcs
$(x, y) \in A$;
    For $l = 1, 2, ..., n - 1$ **do**
        For each subset $\bar{M} \subset A \backslash \{r\}$ such that $|\bar{M}| = l$ **do**
        $M := \bar{M} + \{r\}$;
          For each $x \in V(N[M])$ and $t = 0, 1, ..., T$ **do** $R_z(M, x, t) :=$
∞;
          For each $e = (x, y) \in \bar{M}$ **do** $Q_z(e, M, 0) := \infty$;
          For $t = 1, 2, ..., T$ **do**
              For each $e \in \bar{M}$ **do**

$$Q_z(e, M, t) := \min_{\{u | u + b(x, y, u) = t\}} \{R_z(M \backslash \{e\}, x, u) + c(x, y, u)\};$$

    Call procedure TSP-ZW to calculate all paths from e to
each other vertex $z \in V(N(M))$ and update $R(M, z, t')$;
    Let $z^* := \min_{e \in A} \min_{0 \leq t \leq T} Q_z(e, A, t)$;
    End.

Theorem 8.14 *Algorithm TVCPP-ZW can optimally solve the TVCP problem with the zero waiting constraint in $O(m^2 n T^2 2^m)$ time.*

Definition 8.17 *Let $M \subset A$ and $r \in M$, define $Q_b(e, M, t)$ as the cost of a shortest dynamic Chinese path from r to e of time exactly t in M under the bounded waiting time constraints. If this path dose not exist, let $Q_b(e, M, t) = \infty$.*

Lemma 8.17 *$Q_b(r, M = \{r\}, 0) = 0$ and for any time $0 < t \leq T$, $Q_b(r, M = \{r\}, t) = \infty$. For any arc $e \in M$ and $e \neq r$, if $r \notin M$, $Q_b(e, M, t) = \infty$ for any time t; otherwise, if $r \in M$, $Q_b(e, M, 0) = \infty$ and*

$$Q_b(e, M, t)$$
$$= \min_{g \in M \backslash \{e\}} \min_{(u_A, u_D) \in \mathcal{H}(g, e, t)} \{Q_b(g, M \backslash \{e\}, u_A) + \zeta(P_M(g, e))\}\},$$

where $\mathcal{H}(g, e, t) = \{(u_A, u_D) | \tau(g) = u_D, \alpha(e) = t, 0 \leq u_D - u_A \leq u_x\}$, $P_M(g, e)$ is the dynamic shortest path from g to e in M with the departure time $\tau(g)$ and arrival time $\alpha(e)$ and waiting at vertex x is bounded by a given number u_x.

Definition 8.18 *Let $M \subset A$. Define $R_b(M, x, t)$ as the cost of a shortest dynamic Chinese path in M that ends at vertex x within time t, where waiting at any vertex x is bounded by a given number u_x. If such a path does not exist, let $R_b(M, x, t) = \infty$.*

Lemma 8.18 *For each $M \subset A$, each vertex $x \in V(N[M])$ and each time $0 \leq t \leq T$, we have*

$$R_b(M, x, t)$$

$$= \min_{e=(y,z) \in M} \min_{(u_A, u_D) \in \mathcal{H}(g, e, t)} \{Q_b(e, M, u_A) + \zeta(P_M(z, x))\}.$$

The calculation shown in Lemma 8.17 can be rewritten to

$$Q_b(e, M, t) :=$$

$$\min_{\{(u_A, u_D) | u_D + b(x, y, u_D) = t, 0 \leq u_D - u_A \leq u_x\}} \{R_b(M \backslash \{e\}, x, u_A) + c(x, y, u_D)\},$$

where $e = (x, y) \in M$.

> **Algorithm TVCPP-BW**
> **Begin**
> **Initialize**: $Q_b(r, \{r\}, 0) := 0$ and $Q_b(r, \{r\}, t) := \infty$ for $t > 0$; $Q_b(e, \{r\}, t) := \infty$ for $0 \leq t \leq T$;
> **Sort** all values $u + b(x, y, u) = t$ for $u = 1, 2, ..., T$, and for all arcs $(x, y) \in A$;
> **For** $l = 1, 2, ..., n-1$ **do**
> **For** each subset $\bar{M} \subset A \backslash \{r\}$ such that $|\bar{M}| = l$ **do**
> $M := \bar{M} + \{r\}$;
> **For** each $x \in V(N[M])$ and $t = 0, 1, ..., T$ **do** $R_b(M, x, t) :=$ ∞;
> **For** each $e \in \bar{M}$ **do** $Q_b(e, M, 0) := \infty$;
> **For** $t = 1, 2, ..., T$ **do**
> **For** each $e = (x, y) \in \bar{M}$ **do**
> $Q_b(e, M, t) :=$
> $\min_{\{(u_A, u_D) | u_D + b(x, y, u_D) = t, 0 \leq u_D - u_A \leq u_x\}} \{R_b(M \backslash \{e\}, x, u_A) + c(x, y, u_D)\}$;
> Call procedure TSP-BW to calculate all paths from e to each other vertex $z \in V(N[M])$ and update $R(M, z, t')$;
> **Let** $z^* := \min_{e \in A} \min_t Q_b(e, A, t)$;
> **End.**

Theorem 8.15 *Algorithm TVCPP-BW can optimally solve the TVCP problem with the bounded waiting constraint in $O(T^2 2^m m(m + n \log T))$ time.*

5.   Additional references and comments

The generalized maximum flow problem can be converted to a linear program, and solved using general-purpose linear programming algorithms, such as the ellipsoid method (Karchian (1980)) or Karmarkar's algorithm (Karmarkar (1984)). The fastest of this kind of algorithms currently known runs in $O(m^{3.5} \log(nB))$ time when applied to the generalized maximum flow problem, where B is the largest integer used to represent the multipliers and capacities (Vaidya (1989)). Using the technique of Kapoor et al (1986), this algorithm can be modified to run in $O(n^2 m 1.5 \log(nB))$ time (Vaidya (1989)). Another algorithm for the maximum generalized flow problem is reported by Goldberg, Plolkin and Tardos, which can be implemented in $O(n^2 m^2 \log n \log^2 B)$ (Goldberg (1998)).

A generalized time-expanded network can be created to convert a maximum generalized flow problem with arc transit times to an equivalent static problem without arc transit times, which can then be solved using a method for solving the static problem (Ford et al (1962)). However, the size of this time-expanded network is much greater than that of the original one. Let m and n denote the number of arcs and vertices of the original network, m' and n' the number of arcs and vertices of the equivalent time-expanded network, respectively. Then, if waiting at vertices is arbitrarily allowed, then we have $m' = (m + n)T$ and $n' = nT$. Therefore, if we apply the algorithm of Goldberg (1998), the running time will be $O(T^4 m^2 n^2 \log(nT) \log^2 B)$. The time complexity of the algorithm we have presented in this chapter is bounded above by $O(T^4 m^3 n \log B)$. For the case where all multipliers are less than or equal to 1 and vertex capacities are unbounded, the time complexity of our algorithm can be further reduced to $O(T^3 m^3 n \log B)$.

References

Abdel-Wahab, H., I. Stoica, F. Sultan, K. Wilson. (1997), A simple algorithm for computing minimum spanning trees in the Internet. *Information Sciences*, Vol. 101, pp. 47-69.

Adel'son-Vel'ski, G.M., Dinic, E.A. and Karzanov, E.V. (1975), *Flow Algorithms*, Science, Moscow.

Ahuja, R. K. (1988), Minimum cost-reliability ratio path problem, *Computers and Operations Research*, Vol. 15, No. 1, pp. 83-89.

Ahuja, R. K., J.L. Batra and S.K. Gupta(1983), Combinatorial optimization with rational objective functions: A communication, *Mathematics of Operations Research*, Vol. 8, pp. 314.

Ahuja, R.K. and James, B. Orlin, J.B. (1991), Distance-directed augmenting path algorithms for maximum flow and parametric maximum flow problems, *Naval Research Logistics*, Vol. 38, pp. 413-430.

Ahuja, R.K., Magnanti, T.L. and Orlin, J.B. (1991), Some recent advances in network flows. *SIAM Review*, Vol. 33, pp. 175-219.

Ahuja, R.K., Magnanti, T.L. and Orlin, J.B. (1993), *Network Flows: Theory, Algorithms and Applications*, Prentice Hall, Englewood Cliff, New Jersey.

Anderson, E.J., Nash, P. and Philpott, A.B. (1982), A class of continuous network flow problems, *Mathematics of Operations Research*, Vol. 7, No. 4, pp. 501-514.

Anderson, E.J. and Philpott, A.B. (1984), Duality and algorithm for a class of continuous transportation problems, *Mathematics of Operations Research*, Vol. 9, No. 2, pp. 222-231.

Anderson, E.J. and Philpott, A.B. (1994), *Optimization of flows in networks over time, probability, statistics and optimization (Edited by F. P. Kelly)*, John Wiley & Sons Ltd..

Aneja, Y. P. and Nair, K. P. K. (1978), The constrained shortest path problem, *Naval Research Logistics*, Vol. 25, pp. 549-555.

Aronson, J.E. (1986), The multiperiod assignment problem: a multicommodity network flow model and specialized branch and bound algorithm, *European Journal of Operational Research*, Vol. 23, pp. 367-381.

Aronson, J.E. (1989), A survey of dynamic network flows, *Annals of Operations Research*, Vol. 20, No. 1/4, pp. 1-66.

Aronson, J.E. and Chen, B.D. (1986), A forward network simplex algorithm for solving multiperiod network flow problems, *Naval Research Logistics*, Vol. 33, pp. 445-467.

Aronson, J.E. and Chen, B.D. (1989), A primary/secondary memory implementation of a forward network simplex algorithm for multiperiod network flow problems, *Computers and Operations Research*, Vol. 16, No. 4, pp. 379-391.

Aronson, J.E., Morton, T.E. and Thompson, G.L. (1984a), A forward algorithm and planning horizon procedure for the production smoothing problem without inventory, *European Journal of Operational Research*, Vol. 15, pp. 348-365.

Aronson, J.E., Morton, T.E. and Thompson, G.L. (1985a), A forward simplex method for staircase linear programs, *Management Science*, Vol. 31, No. 6, pp. 664-679.

Aronson, J.E. and Thompson, G.L. (1984b), A survey on forward methods in mathematical programming, *Large Scale Systems*, Vol. 7, pp. 1-16.

Aronson, J.E. and Thompson, G.L. (1985b), The solution of multiperiod personnel planning problems by the forward simplex method, *Large Scale Systems*, Vol 9, pp. 129-139.

Arora, S., Lund, C., Motwani, R., Sudan, M. and Szegedy, M. (1992). Proof verification and hardness of approximation problems. *Proceeding of the 33th Annual IEEE Symp. on Foundations of Computer Science*, pp. 14-23.

Baker, E.K. (1983), An exact algorithm for the time-constrained traveling salesman problem, *Operations Research*, Vol. 31, No. 5, pp. 938-945.

Baker, E.K. and Schaffer, J.R. (1986), Solution improvement heuristics for the vehicle routing and scheduling problem with time window constraints, *American Journal of Mathematical and Management Sciences*, Vol. 6, No. 3 & 4, pp. 261-300.

Bazaraa, M.S., Jarvis, J.J. and Sherali, H.D. (1990), *Linear Programming and Network Flows (2nd ed.)*, Wiley, New York.

Bean, J.C., Birge, J.R. and Smith, R.L. (1987), Aggregation in dynamic programming, *Operations Research*, Vol. 35, No. 2, pp. 215-220.

Bean, J.C., Lohmann, J.R. and Smith, R.L. (1984), A dynamic infinite horizon replacement economy decision model, *The Engineering Economist*, Vol. 30, No. 2, pp. 99-120.

Bellman, R. (1958), On a routing problem, *Quarterly of Applied Mathematics*, Vol. 16, pp. 87-90.

Bellmore, M., Ramakrishna, R. and Vemuganti, R.R. (1973), On multi-commodity maximal dynamic flows, *Operations Research*, Vol. 21, No. 1, pp. 10-21.

Berman, O. and Handler, G. Y.(1987), Optimal minimax path of a single service unit on a network to nonservice destinations, *Information Processing Letters*, Vol. 7, pp. 10-14.

Bertsekas, D.P. (1998), *Network Optimization: Continuous and Discrete Models*, Athena Scientific, Nashua.

Blum, W. (1990), A measure-theoretical max-flow-min-cut problem, *Mathematische Zeitschrift*, Vol. 205, pp. 451-470.

Blum, W. (1993), An approximation for a continuous max-flow problem, *Mathematics of Operations Research*, Vol. 18, No. 1, pp. 98-115.

Bodin, L. Golden, B., Assad, A. and Ball, M. (1983), Routing and scheduling of vehicles and crews: the state of the art, *Computers and Operations Research*, Vol. 10, No. 2, pp. 63-211.

Boldberg, A. V., Plotkin, S. A. and Tardos, E. (1991), Combinatorial algorithms for the generalized circulation problem, *Mathematics of Operations Research*, Vol. 16, No. 2, pp. 351-381.

Brideau, R.J., III, and Cavalier, T.M. (1994), The maximum collection problem with time dependent rewards, presented at the *TIMS International Conference*, Alaska, June.

Burkard, R. E., Dlaska, K. and Kellerer H. (1995), The quickest path disjoint flow problem, *Central European Journal for Operations Research and Economics*, Vol. 3, No. 4, pp. 325-337.

Cai, X., Kloks, T. and Wong, C.K. (1997), Time-varying shortest path problems with constraints, *Networks*, Vol. 29, No. 3, pp. 141-149.

Cai, X., Sha, D. and Wong, C.K. (1998), The Time-varying minimum spanning tree problem with waiting times, In J. F. Gu, G. H. Fan, S. Y. Wang, and B. Wei (eds): *Advances in Operations Research and Systems Engineering*, Global-link Publishing Co., Hong Kong, pp. 1-8.

Cai, X., Sha, D. and Wong, C. K. (1998), The time-varying maximum capacity path problem with no waiting time. In J. F. Gu, G. H. Fan, S. Y. Wang, and B. Wei (eds): *Advances in Operations Research and Systems Engineering*. Global-link Publishing Co., pp. 9-16.

Cai, X., Sha, D. and Wong, C.K. (2001a), Time-varying minimum cost flow problems. *European Journal of Operational Research*, Vol. 131, pp. 352-374.

Cai, X., Sha, D. and Wong, C.K. (2001b), Time-varying universal maximum flow problems. *Mathematical and Computer Modelling*, Vol. 33, pp. 407-430.

Cai, X., Sha, D. and Wong, C.K. (2001c), Time-varying k-shortest path problems with constraints. *working paper*.

Cai, X., Sha, D. and Wong, C.K. (2001d), Time-varying minimum cost-reliability ratio path problems. *working paper*.

Cai, X., Sha, D. and Wong, C.K. (2001e), Time-varying quickest path problems. *working paper*.

Carey, M. (1987), Optimal time-varying flows on congested networks, *Operations Research*, Vol. 35, No. 1, pp. 58-69.

Carey, M. and Srinivasan, A. (1988), Congested network flows: time-varying demands and start-time policies, *European Journal of Operational Research*, Vol. 36, pp. 227-240.

Carey, M. and Srinivasan, A. (1993), Externalities, average and marginal costs, and tolls on congested networks with time-varying flows, *Operations Research*, Vol. 41, No. 1, pp. 217-231.

Carey, M. and Srinivasan, A. (1994), Solving a class of network models for dynamic flow control, *European Journal of Operational Research*, Vol. 75, pp. 151-170.

Carlton, W.B. and Barnes J.W. (1996), Solving the traveling-salesman problem with time windows using tabu search, *IIE Transactions*, Vol. 28, pp. 617-629.

Cattrysse, D.G. and Wassenhove, L.N.V. (1992), A survey of algorithms for the generalized assignment problem, *European Journal of Operational Research*, Vol. 60, pp. 260-272.

Cattrysse, L.G., Francis, R.L. and Saunders, P.B. (1982), Network models for building evacuation, *Management Science*, Vol. 28, No. 1, pp. 86-105.

Chang, H. and Chung, K. (1992), Fault-tolerant routing in unique-path multistage omega network, *Information Processing Letters*, Vol. 44, pp. 201-204.

Chardaire, P., Sutter, A. and Costa, M. (1996), Solving the dynamic facility location problem, *Networks*, Vol. 28, pp. 117-124.

Chari, K. and Dutta, A. (1993a), Design of private backbone networks-I: Time varying traffic, *European Journal of Operational Research*, Vol. 67, pp. 428-442.

Chari, K. and Dutta, A. (1993b), Design of private backbone networks-II: Time varying grouped traffic, *European Journal of Operational Research*, Vol. 67, pp. 443-452.

Chen, Y. L. and Chin, Y. H.(1990), "The Quickest Path Problem," *Computers and Operations Research*, Vol. 17, No. 2, pp. 153-161.

Chen, H. and Yao, D.D. (1993), Dynamic scheduling of a multiclass fluid network, *Operations Research*, Vol. 41, No. 6, pp. 1104-1115.

Cheung, R.K. and Powell, W.B. (1996), An algorithm for multistage dynamic networks with random arc capacities, with an application to dynamic fleet management, *Operations Research*, Vol. 44, No. 6, pp. 951-963.

Chlebus, B.S. and Diks, K. (1996), Broadcasting in synchronous networks with dynamic faults, *Networks*, Vol. 27, pp. 309-318.

Choi, W., Hamacher, H.W. and Tufekci, S. (1988), Modeling of building evacuation problems by network flows with side constraints, *European Journal of Operational Research*, Vol. 35, pp. 98-110.

Cooks, K.L. and Halsey, E. (1966), The shortest route through a network with time-dependent internodal transit times, *Journal of Mathematical Analysis and Applications*, Vol. 14, pp. 493-498.

Cormen, T. H., Leiserson, C. E. and Rivest, R. L. (1990), *Introduction to Algorithms*, The MIT press, Cambridge, Massachusetts.

Current, J. and Marsh, M. (1993), Multiobjective transportation network design and routing problems: taxonomy and annotation, *European Journal of Operational Research*, Vol. 65, pp. 4-19.

Dell, R.F., Batta, R. and Karwan, M.H. (1996), The multiple vehicle TSP with time windows and equity constraints over a multiple day horizon, *Transportation Science*, Vol. 30, No. 2, pp. 120-133.

Denardo, E.V. and Fox, B.L. (1979), Shortest-route methods: 1. reaching, pruning, and buckets, *Operations Research*, Vol. 27, No. 1, pp. 161-186.

Deo, N. and Pang, C. (1984), Shortest-path algorithms: taxononny and annotation, *Networks*, Vol. 14, pp. 275-323.

Desrochers, M., Desrosiers, J. and Solomon, M. (1992), A new optimization algorithm for the vehicle routing problem with time windows, *Operations Research*, Vol. 40, No. 2, pp. 342-354.

Desrochers, M. and Soumis, F. (1988a), A reoptimization algorithm for the shortest path problem with time windows, *European Journal of Operational Research*, Vol. 35, pp. 242-254.

Desrochers, M. and Soumis, F. (1988b), A generalized permanent labelling algorithm for the shortest path problem with time windows, *Information Systems and Operations Research*, Vol. 26, No. 3, pp. 191-212.

Desrosiers, J., Dumas, Y. and Soumis, F. (1986), A dynamic programming solution of the large-scale single-vehicle dial-a-ride problem with time windows, *American Journal of Mathematical and Management Sciences*, Vol. 6, No. 3 & 4, pp. 301-325.

Desrosiers, J., Pelletier, P. and Sonmis, F. (1983), Plus Court Chemin avec Contraintes d'Horaires, *RAIRO Operations Research*, Vol. 17, No. 4, pp. 357-377.

Dijkstra, E. (1959), A note on two problems in connexion with graphs. *Numerische Mathematik*, Vol. 1, pp. 269-271.

Dreyfus, S.E. (1969), An appraisal of some shortest-path algorithms, *Operations Research*, Vol. 17, pp. 395-412.

Dude, R. O. and Hart, P. E. (1973), *Pattern Classification and Science Analysis*, Wiley, New York.

Duffin, R. J. (1965), Topology of series-parallel networks, *Journal of Mathematical Analysis and Applications*, Vol. 10, pp. 303-318.

Dumas, Y., Desrosiers, J., Gelinas, E. and Solomon, M.M. (1995), An optimal algorithm for the traveling salesman problem with time windows, *Operations Research*, Vol. 43, No. 2, pp. 367-371.

Edmonds, J. (1965), Optimum branchings, *Journal of Research of the National Bureau of Standards*, Vol. 71B, pp. 233-340.

Eiselt, H.A., Gendreau, M. and Laporte, G. (1995a), Arc routing problems, part I: the Chinese postman problem, *Operations Research*, Vol. 43, No. 2, pp. 231-242.

Eiselt, H.A., Gendreau, M. and Laporte, G. (1995b), Arc routing problems, part I: the rural postman problem, *Operations Research*, Vol. 43, No. 3, pp. 399-414.

Elam, J., Glover, F. and Klingman, D. (1979), A strongly convergent primal simplex algorithm for gneralized networks, *Mathematics of Operations Research*, Vol. 4, pp. 39-59.

Erkut, E. and Zhang, J. (1996), The maximum collection problem with time-dependent rewards, *Naval Research Logistics*, Vol. 43, pp. 749-763.

Federgruen, A. and Groenevelt, H. (1986), Preemptive scheduling of uniform machines by ordinary network flow techniques, *Management Science*, Vol. 32, No. 3, pp. 341-349.

Fisher, M.L. (1994), Optimal solution of vehicle routing problems using minimum k-trees, *Operations Research*, Vol. 42, No. 4, pp. 626-642.

Fisher, M.L., Jornsten, K.O. and Madsen, O.B.G. (1997), Vehicle routing with time windows: two optimization algorithms, *Operations Research*, Vol. 45, No. 3, pp. 488-492.

Ford, L.R., Jr. and Fulkerson, D.R. (1956), Maximal flow through a network, *Canadian Journal of Mathematics*, Vol. 8, pp. 399-404.

Ford, L.R., Jr. and Fulkerson, D.R. (1958), Constructing maximal dynamic flows from static flows, *Operations Research*, Vol. 6, pp. 419-433.

Ford, L.R., Jr. and Fulkerson, D.R. (1962), *Flows in Networks*, Princeton University Press, Princeton, NJ.

Franca, P.M., Gendreau, M., Laporte, G. and Muller, F.M. (1995), The m-traveling salesman problem with minimax objective, *Transportation Science*, Vol. 29, No. 3, pp. 267-275.

Frederickson, G. N. (1993), An optimal algorithm for selection in a Min-Heap, *Information and Computation*, Vol. 104, pp. 197-214.

Fredman, M. L. and Tarjan, R. E. (1987), Fibonacci heaps and their uses in improved network optimization algorithms, *Journal 0f the ACM*, Vol. 34, pp. 596-615.

Fukuda, K. and Matsul, T. (1992), Finding all minimum-cost perfect matchings in bipartite graphs, *Networks*, Vol. 22, pp. 461-468.

Gabow, H. N.(1985), Scaling algorithms for network problems, *Journal of Computer and System Sciences*, Vol. 31, pp. 148-168.

Gabow, H. N., Galil, Z., Spencer, T. and Tarjan, R.E. (1986), Efficient algorithms for finding minimum spanning trees in undirected and directed graphs, *Combinatorica*, Vol. 6, pp. 109-122.

Gaimon, C. (1986), Optimal inventory, backlogging and machine loading in a serial multi-stage, multi-period production environment, *International Journal of Production Research*, Vol. 24, No. 3, pp. 647-662.

Gale, D. (1957), A theorem on flows in networks, *Pacific Journal of Mathematics*, Vol. 7, pp. 1073-1086.

Gale, D. (1959), Transient flows in network, *The Michigan Mathematical Journal*, Vol. 6, pp. 59-63.

Garey, M. R. and Johnson, D. S. (1979), *Computers and Intractability: A Guide to the Theory of NP-Completeness*. W. H. Freeman and Company, New York.

Gilberg, R. F. and Forouzan, B. A. (2001), *Data structures, A pseudocode approach with c++*, Brooks/Cole, A division of Thomson Learning, CA.

Gilmore, P. C. and Gomory, R. E. (1964), Sequencing a one state-variable machine: A solvable case of the travelling salesman problem, *Operations Research*, Vol. 12, pp. 655-679.

Glover, F. (1990), Tabu search: a tutorial, *Interfaces*, Vol. 20, pp. 74-94.

Glover, F., Hultz, J., Klingman, D. and Stutz, J. (1978), Generalized network: a fundamental computer-based planning tool, *Management Science*, Vol. 24, pp. 12.

Glover, F. and Klingman, D.(1973), On the equivalence of some generalized network problem to pure network problems, *Mathematical Programming*, Vol. 4, pp. 269-278.

Glover, F. and Laguna, M. (1997), *Tabu Search*, Kluwer Academic Publishers, Norwell, MA.

Goczyla, K. and Cielatkowski, J. (1995), Optimal routing in a transportation network, *European Journal of Operational Research*, Vol. 87, pp. 214-222.

Goldberg, A.V. (1998), Recent developments in maximum flow algorithms, *proceedings of Sixth Scandinavian Workshop on Algorithm*.

Gondran, M. and Minoux, M. (1984), *Graphs and Algorithms*, Wiley-Interscience, New York.

Gower, J. C. and Ross, G. J. S. (1969), Minimum spanning trees and single linkage cluster analysis, *Applied Statistics*, Vol. 18, pp. 54-64.

Graham, R. L. and Hell, P. (1985), On the history of minimum spanning tree problem, *Annals of the History of Computing*, Vol. 7, pp. 43-57.

Grossmann, W., Guariso, G., Hitz, M. and Werthner, H. (1995), A min cost flow solution for dynamic assignment problems in Networks with storage devices, *Management Science*, Vol. 41, No. 1, pp. 83-93.

Gupta, S.K. (1985), *Linear Programming and Network Models*. Affiliated East-West Press, New Delhi, India.

Halpern, I. (1979), Generalized dynamic flows, *Networks*, Vol. 9, pp. 133-167.

Hamacher, H.W. and Tufekci, S. (1987), On the use of lexicographic min cost flows in evacuation modeling, *Naval Research Logistics*, Vol. 34, pp. 487-503.

Handler, G. Y. and Zang, I. (1980), A dual algorithm for the constrained shortest path proglem, *Networks*, Vol. 10, pp. 293-310.

Hansen, P.(1980), Bicriterion Path Problems, *Lecture Notes in Economics and Mathematical Systems*, No. 177, Springer Verlag, Berlin, pp. 109-127.

Hassan, M. M. D. (1992), Network reduction for the acyclic constrained shortest path problem, *European Journal of Operational Research*, Vol. 63, pp. 121-132.

Hiroshi, M. (1993). A New kth-Shortest Path Algorithm, *IEICE Transactions on Information and Systems*, Vol. 76, pp. 388-389.

Ichimori, T., Ishi, H. and Nishida, T.(1979), A minimax flow problem, *Mathematica Japonica*, Vol. 24, pp. 65-71.

Imase, M. and Waxman, B.M. (1991), Dynamic steiner tree problem, *SIAM Journal on Discrete Mathematics*, Vol. 4, No. 3, pp. 369-384.

Ioachim, I., Gelinas, S., Francois, and Desrosiers, J. (1998), A dynamic programming algorithm for the shortest path problem with time windows and linear node costs, *Networks*, Vol. 31, pp. 193-204.

Iri, M. (1969), *Network Flow, Transportation and Scheduling*, Academic Press, New York.

Jacobs, K. and Seiffert, G. (1983), A measure-theoretical max-flow problem, Part I, *Bulletin of the Institute of Mathematics. Academia Sinica*, Vol 11, No. 2, pp. 261-280.

Jaillet, P. (1992), Shortest path problems with node failures, *Networks*, Vol. 22, pp. 589-605.

Janson, B.N. (1991), Dynamic traffic assignment for urban road networks, *Transportation Research Part B*, Vol. 25, No. 2/3, pp. 143-161.

Jarvis, J.J. and Ratliff, H.D. (1982), Some equivalent objectives for dynamic network flow problems, *Management Science*, Vol. 28, No. 1, pp. 106-109.

Jensen, P.A. and Barnes, W. (1980), *Network Flow Programming*, Wiley, New York.

Joksch, H. C. (1966), The shortest route probelm with constraints, *Journal of Mathematical Analysis and Application*, Vol. 14, pp. 191-197.

Jordan, W. C. and Turnquist, M. A.(1983), A stochastic, dynamic network model for railroad car distribution, *Transportation Science*, Vol. 17, No. 2, pp. 123-145.

Kang, A. N. C., Lee, R. C. T., Chang, C. L. and Chang, S. K. (1977), Storage reduction through minimal spanning trees and spanning forests, *IEEE Transactions on Computers*, Vol. 26, pp. 425-434.

Kappor, S. and Vaidya, P. M.(1986), Fast algorithms for convex quadratic programming and multicommodity flows, In *Proc. 18th Annual ACM Symposium on Theory of Computing*, pp. 147-159.

Karmarkar, N.(1984), A new polynomial-time algorithm for linear programming, *Combinatorica*, Vol. 4, pp. 373-395.

Khachian, L. G.(1980), Polynomial algorithms in linear programming, *Zhurnal Vychislitelnoi Matematiki i Matematicheskoi Fiziki*, Vol. 20, pp. 53-72.

Klingman, D. and Mote, J. (1982), A multi-period production, distribution, and inventory planning model, *Advances in Management Studies*, Vol. 1, pp. 56-76.

Kogan, K., Shtub, A. and Levit, V.E. (1997), DGAP-the dynamic generalized assignment problem, *Annals of Operations Research*, Vol. 69, pp. 227-239.

Kruskal, J. (1956), On the shortest spanning subtree of a graph and the travelling salesman problem. *Proceedings of the American Mathematical Society*, Vol. 7, pp. 48-50.

Lawler, E. L. (1972), A procedure for computing the K best solutions to discrete optimization problems and its application to the shortet path problem, *Management Science*, Vol. 18, pp. 401-405.

Lawler, E.L. (1976), *Combinational Optimization: Networks and Matroids*, Holt Rinehart and Winston, New York.

Lee, D.T. and Papadopoulou, E. (1993), The all-pairs quickest path problem, *Information Processing Letters*, Vol. 45, pp. 261-267.

Lee, H. and Pulat, P.S. (1991), Bicriteria network flow problems: continuous case, *European Journal of Operational Research*, Vol. 51, pp. 119-126.

Levner, E.V. and Nemirovsky, A.S. (1994), A network flow algorithm for just-in-time project scheduling, *European Journal of Operational Research*, Vol. 79, pp. 167-175.

Liu, P. C., Geldmacher, R. C. (1976), Graph reducibility, *Proceedings, Seventh Southeastern Conference on Combinatorics, Graph Theory, and Computing*, pp. 433-445.

Liu, P. C., Geldmacher, R. C. (1980), An $O(max(m, n))$ algorithm for finding a subgraph homomorphic to K_4. *Proceedings, Eleventh Southeastern Conference on Combinatorics, Graph Theory, and Computing*, pp. 597-609.

Loberman, H. and Weinberger, A. (1957), Formal procedures for connecting terminals with a minimum total wire length, *Journal of the ACM*, Vol. 4, pp. 428-437.

Lund, C. and Yannakakis, M. (1993), On the hardness of approximation problems, *Proceeding of the 25th Annual ACM Symp. on Theory of Computing*, pp. 14-23.

Magnanti, T. L. and Wong, R. T. (1984), Network design and trans, *Transportation Science*, Vol. 18, pp. 1-55.

Martins, E. Q. V. and Santos, J. L. E. (1997), An algorithm for the quickest path problem, *Operations Research Letters*, Vol. 20, pp. 195-198.

Mcquillan, J.M., Richer, I. and Rosen, E.C. (1980), The new routing algorithm for the ARPANET, *IEEE Transactions on Communications*, Vol. 28, No. 5, pp. 711-719.

Merchant, D.K. and Nemhauser, G.L. (1978a), A model and an algorithm for the dynamic traffic assignment problems, *Transportation Science*, Vol. 12, No. 3, pp. 183-199.

Merchant, D.K. and Nemhauser, G.L. (1978b), Optimality conditions for a dynamic traffic assignment model, *Transportation Science*, Vol. 12, No. 3, pp. 200-207.

Miller-Hooks, E. D.(2001), Adaptive Least-expected time paths in stochastic, time-varying transportation and data networks, *Networks*, Vol. 37, No. 1, pp. 35-52.

Miller-Hooks, E. D. and Mahmassani, H. S.(2000), Least expected time paths in stochastic, time-varying transportation networks, *Transportation Science*, Vol. 34, No. 2, pp. 198-215.

Mingozzi, A., Bianco, L. and Ricciardelli, S. (1997), Dynamic programming strategies for the traveling salesman problem with time window and precedence constraints, *Operations Research*, Vol. 45, No. 3, pp. 365-377.

Minieka, E. (1973), Maximal, lexicographic, and dynamic network flows, *Operations Research*, Vol. 12, No. 2, pp. 517-527.

Minieka, E. (1978), *Optimization Algorithms for Networks and Graphs*, Marcel Dekker, New York.

Mote, J., Murthy, I. and Olson, D. L. (1991), A Parametric Approach to Solveing Bicriterion Shortest Path Problem, *European Journal of Operational Research*, Vol. 53, pp. 81-92.

Mulvey, J. M. and Vladimirou, H.(1992), Stohastic network programming for Financial Planning Problems, *management Science*, Vol. 38, No. 11, pp. 1642-1664.

Murthy, I. (1993), Solving the multiperiod assignment problem with start-up cost using dual ascent, *Naval Research Logistics*, Vol. 40, pp. 325-344.

Nachtigall, K. (1995), Time depending shortest-path problems with applications to railway networks, *European Journal of Operational Research*, Vol. 83, pp. 154-166.

Nanry, W.P. and Barnes, J.W. (2000), Solving the pickup and delivery problem with time windows using reactive tabu search, *Transportation Research Part B*, Vol. 34, pp. 107-121.

Nozawa, R. (1994), Examples of max-flow and min-cut problems with duality gaps in continuous networks, *Mathematical Programming*, Vol. 63, pp. 213-234.

Orda, A. and Rom, R. (1990), Shortest-path and minimum-delay algorithms in networks with time-dependent edge-length, *Journal of Association for Computing Machinery*, Vol. 37, No. 3, pp. 607-625.

Orda, A. and Rom, R. (1991), Minimum weight paths in time-dependent networks, *Networks*, Vol. 21, pp. 295-319.

Orda, A., Rom, R. and Sidi, M.(1993), Minimum delay routing in stochastic networks, *IEEE/ACM Transactions on Networking*, Vol. 1, No. 2, pp. 187-198.

Orda, A. and Rom, R. (1995), On continuous network flows, *Operations Research Letters*, Vol. 17, pp. 27-36.

Orlin, J.B. (1983), Maximum-throughput dynamic network flows, *Mathematical Programming*, Vol. 27, No. 2, pp. 214-231.

Orlin, J.B. (1984), Minimum convex cost dynamic network flows, *Mathematics of Operations Research*, Vol. 9, No. 2, pp. 190-207.

Osteen, R. E. and Lin, P. P. (1974), Picture skeletons based on eccentricities of points of minimum spanning trees, *SIAM Journal on Computing*, Vol. 3, pp. 23-40.

Papadimitrion, C. and Steiglitz, K. (1982), *Combinatorial Optimization: Algorithms and Complexity*, Prentice-Hall, Endlewood Cliff, NJ.

Papadimitrion, C. H. and Yannakakis, M. (1991), Optimization, approximation, and complexity classes, *Journal of Computer and System Sciences*, Vol. 43, pp. 425-440.

Papageorgiou, M. (1990), Dynamic modeling, assignment, and route guidance in traffic networks, *Transportation Research Part B*, Vol. 24, No. 6, pp. 471-495.

Philpott, A.B. (1990), Continuous-time flows in networks, *Mathematics of Operations Research*, Vol. 15, No. 4, pp. 640-661.

Philpott, A.B. and Mees, A.I. (1992), Continuous-time shortest path problems with stopping and starting costs, *Applied Mathematics Letters*, Vol. 5, No. 5, pp. 63-66.

Picard, J. and Queyranne M. (1978), The time-dependent traveling salesman problem and its application to the tardiness problem in one-machine scheduling, *Operations Research*, Vol. 26, No. 1, pp. 86-110.

Polak, G.G. (1992), On a parametric shortest path problem from primal-dual multicommodity network optimization, *Networks*, Vol. 22, pp. 283-295.

Posner, M.E. and Szwarc, W. (1983), A transportation type aggregate production model with backordering, *Management Science*, Vol. 29, No. 2, pp. 188-199.

Prim, R. C. (1957), Shortest connection networks and some generalizations, *Bell System Technical Journal*, Vol. 36, pp. 1389-1401.

Psaraftis, H.N., Solomon, M.M., Magnanti T.L. and Kim, T. (1990), Routing and scheduling on a shoreline with release times, *Management Science*, Vol. 36, No. 2, pp. 212-223.

Psaraftis, H.N. and Tsitsiklis, J.N. (1993), Dynamic shortest paths in acyclic networks with markovian arc costs, *Operations Research*, Vol. 41, No. 1, pp. 91-101.

Punuen, A. P.(1991), A Linear Algorithm for the Maximum Capacity Path Problem, *European Journal of Operational Research*, Vol. 53, pp. 402-404.

Recski, A. (1988), *Matroid Theory and Its Applications*, Springer-Verlag, New York.

Rego, C. and Roucairol, C. (1995), Using tabu search for solving a dynamic multi-terminal truck dispatching problem, *European Journal of Operational Research*, Vol. 83, pp. 411-429.

Rosen, J. B., Sun, S. Z. and Xue, G. L.(1991), Algorithms for the Quickest Path Problem and the Enumeration of Quickest Paths, *Computers and Operations Research*, Vol. 18, No. 6, pp. 579-584.

Salomon, M., Solomom, M.M., Wassenhove, L.N.V., Dumas, Y. and Dauzere-Peres, S. (1997), Solving the discrete lotsizing and scheduling problem with sequence dependent set-up costs and set-up times using the travelling salesman problem with time windows, *European Journal of Operational Research*, Vol. 100, pp. 494-513.

Sancho, N.G.F. (1992), A dynamic programming solution of a shortest path problem with time constraints on movement and parking, *Journal of Mathematics and Applications*, Vol. 166, pp. 192-198.

Sancho, N.G.F. (1994), Shortest path problem with time windows on nodes and arcs, *Journal of Mathematical Analysis and Applications*, Vol. 186, pp. 643-648.

Saniee, I. (1995), An efficient algorithm for the multiperiod capacity expansion of one location in telecommunications, *Operations Research*, Vol. 43, No. 1, pp. 187-190.

Sexton, T.R. and Choi Y. (1986), Pickup and delivery of partial loads with "soft" time windows, *American Journal of Mathematical and Management Sciences*, Vol. 6, No. 3 & 4, pp. 369-398.

Skiscim, C. C. and Golden, B. L. (1989), Solving k-shortest and constrained shortest path problem efficiently, *Annals of Operations Research*, Vol. 20, pp. 249-282.

Sha, D., Cai, X. and Wong, C.K. (2000), The maximum flow in a time-varying network, *Optimization, Lecture Notes in Economics and Mathematical Systems, 481*, Edited by Nguyen, V. H., Strodiot, J. -J. and Tossings, P., Springer-Verlag Berlin Heidelberg, pp. 437-456.

Shannon, C. E. (1938), A symbolic analysis of relay and switching circuits, *Transactions of the American Institute of Electrical Engineers*, Vol. 57, pp. 713-723.

Sherali, H.D., Ozbay, K. and Subramanian, S. (1998), The time-dependent shortest pair of disjoint paths problem: complexity, models, and algorithms, *Networks*, Vol. 31, pp. 259-272.

Shier, D. (1979), On Algorithms for Finding the k-Shortest Paths in a Network, *Networks*, Vol. 9, pp. 195.

Skiscim, C. C. and Golden, B. L. (1989), Solving k-Shortest and Constrained Shortest Path Problems Efficiently, *Annals of Operations Research*, Vol. 20, pp. 249-282.

Solomom, M.M. (1986), The minimum spanning tree problem with time window constraints, *America Journal of Mathematical and Management Sciences*, Vol. 6, No. 3 & 4, pp. 399-421.

Solomon, M.M. (1987), Algorithms for the vehicle routing and scheduling problems with time window constraints, *Operations Research*, Vol. 35, No. 2, pp. 254-273.

Stanley, J.D. (1987), A forward convex-simplex method, *European Journal of Operational Research*, Vol.29, pp. 328-335.

Steinberg, E. and Napier, H.A. (1980), Optimal multi-level lot sizing for requirements planning systems, *Management Science*, Vol. 26, No. 12, pp. 1258-1271.

Stillinger, F. H. (1967), Physical clusters, surface tension, and critical phenomenon, *Journal of Chemical Physics*, Vol. 47, pp. 2513-2533.

Thangiah, S.R., Potvin, J. and Sun, T. (1996), Heuristic approaches to vehicle routing with backhauls and time windows, *Computers and Operations Research*, Vol. 23, No. 11, pp. 1043-1057.

Truemper, K. (1977), On max flows with gains and pure min-cost flows, *SIAM Journal of Applied mathematics*, Vol. 32, pp. 450-456.

Vaidya, P. M. (1989), Speeding up linear programming using fast matrix multiplication, In *Proc. 30th IEEE Annual Symposium on Foundations of Computer Science*, pp. 332-337.

Valdes, J., Tarjan, R. E. and Lawler, E. (1982), The recognition of series-parallel digraphs, *SIAM Journal on Computing*, Vol. 11, pp. 298-313.

Van Slyke, R. and Frank, H. (1972), Network reliability analysis: Part I, *Networks*, Vol. 1, pp. 279-290.

Van Der Bruggen, L.J.J., Lenstra, J.K. and Schuur, P.C. (1993), Variable-depth search for the single-vehicle pickup and delivery problem with time windows, *Transportation Science*, Vol. 27, No. 3, pp. 298-311.

Vemuganti, R.R., Oblak, M. and Aggarrwal, A. (1989), Network models for fleet management, *Decision Science*, Vol. 20, pp. 182-197.

Vythoulkas, P.C. (1990), A dynamic stochastic assignment model for the analysis of general networks, *Transportation Research*, Vol. 24B, No. 6, pp. 453-469.

Wagner, D. K. and Wan, H. (1993), A polynomial-time simplex method for the maximum k-flow problem, *Mathematical Programming*, Vol. 60, pp. 115-123.

Weinberg, L. (1971), Linear graphs: theorems, algorithms, and applications. In R. E. Kalman, N. DeClaris (eds): *Aspects of Network and System Theory*. Holt, Rinehart, and Winston, N.Y.

White, W. W. and Bomberault, A. M. (1969), A network algorithm for empty freight car allocations, *IBM Systems Journal*, Vol. 8, pp. 147-169.

Wie, B., Tobin, R.L., Friesz, T.L. and Bernstein, D. (1995), A discrete time, nested cost operator approach to the dynamic network user equilibrium problem, *Transportation Science*, Vol. 29, No. 1, pp. 79-92.

Wiel, R.J.V. and Sahinidis, N.V. (1995), Heuristic bounds and test problem generation for the time-dependent traveling salesman problem, *Transportation Science*, Vol. 29, No. 2, pp. 167-183.

Wiel, R.J.V. and Sahinidis, N.V. (1996), An exact solution approach for the time-dependent traveling-salesman problem, *Naval Research Logistics*, Vol. 43, pp. 797-820.

Wilkinson, W.L. (1971), An algorithm for universal maximal dynamic flows in a network, *Operations Research*, Vol. 19, No. 7, pp. 1602-1612.

Witzgall, C. and Goldman, A. J. (1965), Most profitable routing before maintenance, *Bulletin of the Operations Research Society of America*, Vol. B82.

Xu, Y. and Uberbacher, E. C. (1997), 2D image segmentation using minimum spanning trees, *Image and Vision Computing*, Vol. 15, pp. 47-57.

Xue, G., Sun, S. and Rosen, J.B. (1998), Fast data transmission and maximal dynamic flow, *Information Processing Letters*, Vol. 66, pp. 127-132.

Ye, Y. (1997), *Interior Piont Algorithms: Theory and Analysis*, John Wiley & Sons, Inc., New York.

Zahn, C. T. (1971), Graph-theoretical methods for detecting and describing gestalt clusters, *IEEE Transactions on Computing*, Vol. C-20, pp. 68-86.

Zahorik, A., Thomas, L.J. and Trigeiro, W.W. (1984), Network programming models for production scheduling in multi-stage, multi-item capacitated systems, *Management Science*, Vol. 30, No. 3, pp. 308-325.

Zawack, D.J. and Thompson, G.L. (1987), A dynamic space-time network flow model for city traffic congestion, *Transportation Science*, Vol. 21, No. 3, pp. 153-162.

INDEX

Early Titles in the
INTERNATIONAL SERIES IN
OPERATIONS RESEARCH & MANAGEMENT SCIENCE
Frederick S. Hillier, Series Editor, *Stanford University*

* A list of the more recent publications in the series is at the front of the book *

Printed in the United States of America